2012 AND THE GALACTIC CENTER

"2012 and the Galactic Center *engages us in the most exciting event in 260 centuries! Christine Page presents positive challenges to conventional doubts about the future.*"

C. Norman Shealy, coauthor of *Soul Medicine: Awakening Your Inner Blueprint for Abundant Health and Energy*

"Christine Page is a remarkable wisdomkeeper, and her book is a masterful and enlightening synthesis of ageless wisdom that speaks to our troubled times."

Brian Luke Seaward, author of *Stand Like Mountain, Flow Like Water*

2012 AND THE GALACTIC CENTER

THE RETURN OF THE GREAT MOTHER

CHRISTINE R. PAGE, M.D.

Bear & Company
Rochester, Vermont

Bear & Company
One Park Street
Rochester, Vermont 05767
www.BearandCompanyBooks.com

Bear & Company is a division of Inner Traditions International

Library of Congress Cataloging-in-Publication Data
Page, Christine R.
2012 and the galactic center : the return of the great mother / Christine R. Page.
p. cm.
Includes bibliographical references and index.
ISBN 978-1-59143-086-5
1. Prophecies (Occultism) 2. Mayas—Prophecies. I. Title. II. Title: Twenty twelve and the galactic center.
BF1791.P34 2008
299.7'842—dc22
2008016648

Printed and bound in the United States by Lake Book Manufacturing

10 9 8 7 6 5 4 3 2 1

Text design Diana April and layout by Jon Desautels
This book was typeset in Garamond Premier Pro with Omni used as a display typeface

The illustrations on pages 3, 8, 41, 53, 85, 96, 119, and 204 are by Sherrie Frank.
The illustrations on pages 25, 26, 31, 62, 93, 105, 106, 107, 120, 134, 143, 146, 153, 179, 186, 190, and 191 are © 2006, Jupiter Images.
The photograph on page 10 is courtesy of Lucy Pringle.
The photograph on page 83 is courtesy of Robert Temple.
The images on pages 92 and 126 are courtesy of Wikimedia Commons, GNU Free Document License.

For more information about Christine R. Page and her work visit:
www.christinepage.com

CONTENTS

Introduction: Teachings beneath the Stars 1

1 The Creation Myth 21

2 Rhythms of the Moon 40

3 The Hero's Journey 68

4 The Celestial Design 75

5 The Radiant Blueprint 84

6 From Boy-Child to King 96

7 And Then One Day . . . 118

8 The Love that Knows No End 137

9 The Descent 150

10 The Truth Shall Set Them Free 174

11 The Emerald Tablet 188

12 The Lunation Phases and the Nodes of the Moon 203

Conclusion: The Heart of the Great Mother 215

Notes 220

Bibliography 225

Index 227

To the Great Mother who nurtures my soul

and

to my dear husband, Leland,

for his commitment to respect, honor, and

support my life's work

and

to love the goddess within me.

INTRODUCTION

TEACHINGS BENEATH THE STARS

Once again I find myself kneeling in the warm sand, watching my beloved teacher whose face is lined and darkened by years of living in the harsh sun of the desert. It is all so familiar: the hot earth, the exquisite colors of the setting sun, and the trust I have in this man whose age is indeterminate but who carries an air both ancient and eternal.

Our stance is natural for nomadic people: one leg poised, ready to straighten and propel the body into flight at a moment's notice, and the other tucked under the body, allowing rest. It reminds me of the phrase "in the world but not of it." At this moment, my attention is focused on a fairly intricate mandala my teacher has been carefully sculpting in the sand for the past hour. He looks up, pleased with his efforts, and then, without a word and with one swift move of his hand, he sweeps away the whole picture, until all that is left is the smooth, virginal sand.

He smiles. "Remember that despite appearances, life is impermanent; here today gone tomorrow. All dreams and ideas arise from a primal source often described as the Great Mother or ocean of possibilities, and it is to here that our creations will return eventually to fulfill their destiny. What image of your future do you wish me to create now?" He has a glint in his eye, and his hand is positioned over the sand, ready to draw.

"But change can't be as easy as merely sweeping your hand through the old and creating something new," I say, amazed at the simplicity of this idea.

He smiles at my realization. "Of course it is, it always has been. As an expression of the One Mind, we are first and foremost creators and transformers of reality. Unfortunately, many fail to appreciate the realm of possibilities readily available to them, preferring to live within the illusionary state of security offered by a familiar experience that has been well tried and tested. Such individuals strive to rise above scarcity and into abundance without realizing that the poverty they wish to leave behind exists primarily within their own minds."

Satisfied with his demonstration, he settles back on his heel and continues: "The earth and its inhabitants are presently engaged in a powerful time of transformation when the old is giving way to the new. In other words, we are dissolving, not evolving.

"According to the Maya, the 1980s saw the beginning of an extraordinary thirty-six-year journey for the earth and its inhabitants, which reaches its conclusion just before 2020.[1] For the first time in twenty-six thousand years, the sun is most closely aligned with what is known as the Great Cleft, the Dark Rift, or the Black Road of the Milky Way.[2] This road leads directly to the Galactic Center, or heart of the Great

The Dark Rift of the Milky Way

Mother, and it is through this portal that we will gain access to the eternal source of all existence, the Mother herself.[3] We are uniquely positioned here on the earth to travel this road metaphorically and enter the black hole at the center of the galaxy. Here, we will experience the fullness of our potentiality, the unlimited realm of possibilities, and come to know the true meaning of immortality.

"After a period of time that some will perceive as chaos and others as a blissful opportunity, we will leave the Great Mother via a white hole and give birth to dreams and visions, which will become the reality for the many generations that follow.

"As an analogy, this thirty-six-year period of time could be seen to represent the three days when the moon is invisible in our night sky. During this time the old moon 'dies' and the new moon is 'born.' To the Maya, who were master astronomers, this 'dark period' encompasses the end of the world of their fourth sun and the birth of their fifth world at sunrise on December 21, 2012.[4] Their calendar, which has become so well known, covers the final and fifth phase of their fourth world and ends at this particular solstice after 5,125 years. Some have translated this to mean that humankind is in the final days of the end of the world. I, however, believe it is only the beginning."

He stops as we both reflect on the years ahead. In a moment, he continues. "To many people, this in-between time of transition will

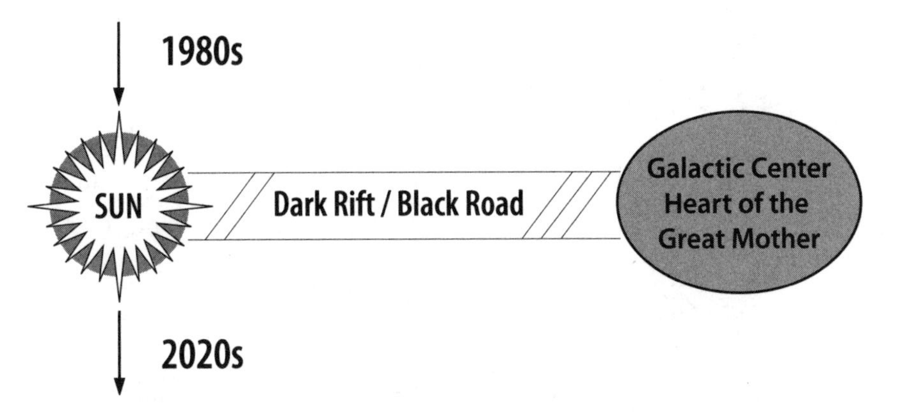

The thirty-six-year passage of the sun across the opening of the Galactic Center.

appear unstructured and confusing, with the past practically complete and the future still to be born. And yet, these extremely unique conditions offer us the ultimate prize that all spiritual beings seek: the opportunity to know eternal life, to become as gods and goddesses.

"For here, between the worlds, time collapses, space expands into infinity, and the veils between the dimensions drop, revealing a single feature: Now our challenge is to be able to release our hold on the past and the future so that we can fully appreciate the rich choices and spiritual promise being offered. Certain aspects are already in place, but the rest will depend on the energy and beliefs we take to the 'cookie shop of opportunities,' the Great Mother, for if all you know are chocolate chip cookies, then that is all you will select from the ocean of possibilities available to you.

"Let me be clear. Immortality is not merely the chance to enter the Great Mother's treasure trove but the ability to move with ease between the worlds of essence and form through focused attention, bringing spirit into matter and then, with a simple change of focus, allowing matter to dissolve back into spirit. The energy required to build the magician's wand is available to us today in a way that it hasn't been for twenty-six thousand years. For now, all of our creations from this immense span of time are returning to us so that we can create a pillar of light, our wand.

"This is not a time to be limited by the fear of change or the unknown or by holding onto a belief that somehow we don't deserve what the Great Mother has to offer. Why do you think there are so many souls on this planet at this time? For lifetimes, everybody has been working toward this moment, determined not to allow this unique opportunity for soul transformation to pass them by, forcing them to wait another twenty-six thousand years."

I let out a long, deep breath, for I know within my being the truth of what he is saying.

He smiles at my recognition. "This definition of immortality—a state of being that recognizes the impermanent and unified field of reality—equates to the teachings of the Maya. These ancient people speak of the earth and its inhabitants entering a new world era ruled by the element of ether, a world where there is a sacred marriage of opposites that will create a unified field of consciousness.

"Ether is considered to be the fifth element and the synthesis of the four other elements: earth, air, fire, and water.[5] According to the great mathematicians—the Pythagoreans—each element is represented by a different geometric form, together known as the Platonic solids. They believed that the constant movement and interrelationship between these elements, each of which can be expressed as a musical frequency, led to the formation of the galaxies and the universes, and life as we know it.

"This new domain of ether is represented by unification, spaciousness, and invisibility and is symbolized by the twelve-faced dodecahedron, which some say is the shape of the universe.[6] The number 12 is held in high regard by religious scholars and mystics. There were twelve tribes of Israel, twelve sons of Jacob, twelve fruits on the Tree of Life, twelve disciples of Jesus, twelve knights at Arthur's Round Table, and there are twelve signs of the zodiac.

"Each time the number is used, we are reminded of the twelve attributes that must be awakened within the human soul before we can gain access to the multidimensional realms of universal awareness, or what some call heaven. I think you have met this shape before?" he asks.

Indeed I had! While studying the perception of reality, I learned that the earth, similar to every human being, possesses an aura consisting of subtle energy bodies or grids of vibratory energy. Each grid

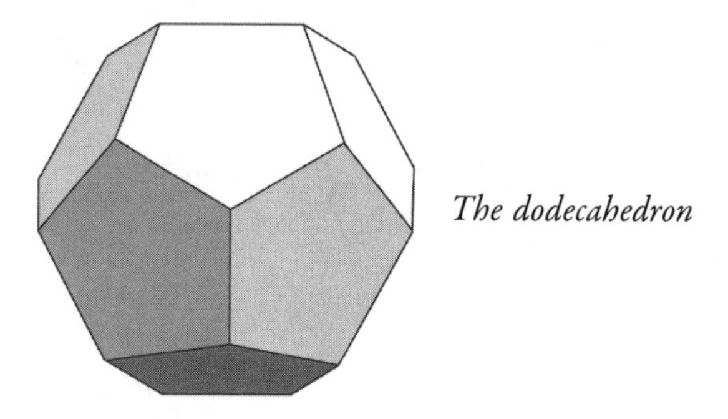

The dodecahedron

is shaped like one of the Platonic solids and fits neatly inside another grid, kept alive by different frequencies of consciousness including our thoughts, feelings, and insights. As any one grid receives more attention and hence power from our collective consciousness, it emits a holographic image, which we then consider to be reality. Hence, when the majority of the population expresses joy, we create and see the world as joyful. When fear is the primary emotion, the world becomes a place of secrets and anxiety exacerbated by our survival responses.

Over time we have come to forget the illusionary nature of the holographic images held in place by our own conscious expression and believe our world is solid and outside our control. Yet in essence, these collections of vibrating energy are capable of transformation as quickly as it takes for us to change our minds!

For thousands of years, the driving forces for the creation of reality have been our emotions, linked to the element of water, and our beliefs and reason, representing the element of air. Now, at the dawning of the new age of compassion, harmony, and interrelatedness, we are reconnecting to a grid that has been patiently awaiting our remembering. This is the unity or Christ grid, which is shaped like a dodecahedron and reflects the element of ether.

He nods as I see clearly how everything ties together. Then he continues: "Ether, the element of the fifth sun, is celestial and lacks material substance, yet it is no less real than wood, wind, flame, stone, or flesh.[7] Within the context of ether there can be a fusion of polarities without the need for separation into darkness and light, negative and positive, good and bad, or even spirit and matter. All are accepted as just another expression of the same essence, lovingly created in the presence of the Great Mother.

"As the poles of existence start to fuse and the veils between the dimensions dissolve, we will find ourselves becoming increasingly sensitive to the effect of our actions on others.[8] In other words, through our hearts we will feel what others feel without having the luxury of guilt, blame, denial, and projection. In this way, we will experience the true meaning of a sentiment found in many religions, including Buddhism, which states: 'Hurt not others in ways that you yourself would find hurtful,' for you and I are one.

"Imagine a world where denial of our interrelatedness is no longer an option. Imagine how a decision to injure or abuse others, whether physically or emotionally, would alter if we knew that their pain would be our pain. When you and I are one, all feelings are shared, and instinctively we will select those actions that bring the greatest degree of harmony and joy. In essence, the fifth world of ether presents us with the possibility of a peace upon this planet that truly passes all our present-day understanding."

We both look out across the desert as each of us reflects on a world where true harmonic unification occurs through the acceptance and celebration of diversity.

He resumes: "There are, of course, some who would prefer to continue the enmity that exists between the different poles of reality in order to feed their own selfish needs. They use fear, distrust, and shame to keep people separate, knowing that cooperation, compassion, and trust would lead to a disintegration of their authority. They are masters of charmed threats, which emotionally persuade others not to rock the boat and distort the face of unity by saying 'my way or the highway.' To perpetuate their own cause, such individuals will exploit the chaos and uncertainty that is common and naturally present at a time like this."

It is not hard for me to identify the individuals or groups who have a vested interest in maintaining control through the incitement of fear. "Yet despite the propaganda," he continues, "large numbers of ordinary people are awakening to the fact that many of the dogmas that have been taught for centuries are based on flawed reasoning. Millions of people are beginning to listen to their own inherent guidance, which tells a very different tale. Through inner knowing or intuition, they are reconnecting to the pulse of the Great Mother's heart and remembering that their destiny in this life is to seed the consciousness of the new world.

"There are those, however, who are preoccupied with shame from the past or fear of the future and who prefer to follow the old, familiar ways. Yet their cellular intelligence is listening, and even though the truth is dormant, it is never lost. During the decades to come, there will be a great need for all who consciously hear the call to encourage others to remember their inherent destiny and hence birth a world that encompasses the principles of cooperation and unity through the acceptance of diversity and right relationship with all who share this planet."

He pauses to allow me to appreciate both the challenges and joys that lie ahead and then looks out into the endless sky. "For those ancient peoples who believed that their existence stretched beyond the confines of the earth plane, the Milky Way was seen as the source of all creation," he continues. "Some cultures saw it as the Great Mother, her heart located at the center, the expansive white area of stars symbolizing her pregnant womb and the Dark Rift representing her birth cleft or vagina.[9]

"For others, our galaxy was a great serpent, its gaping mouth represented by the Dark Rift dividing the white river of stars.[10] Whatever the culture or analogy, the story is the same. It is from there that we are born and nurtured, and it will be there that we will return to die."

He stops while we both turn to honor the passing of the sun as it dips beneath the horizon for another day. "The Maya kings underwent shamanic journeys in service to their people," he continues. "Through ritualistic practices, they entered the serpent's mouth—the portal to the underworld—and traveled through the darkness until they reached

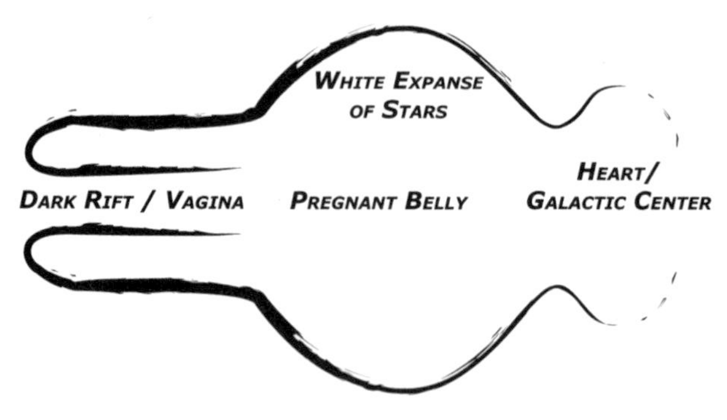

The Milky Way as the Great Mother

the eternal source of all creation.[11] Having immersed themselves in this ocean of possibilities, the consciousness of the Great Mother, they were reborn from that same orifice, bringing wisdom and inspiration to their followers. As you can imagine, the ability of these oracles to move and commune between the worlds ensured that they had an extremely powerful position in their society.

"What is important for you to understand is that rather than merely sharing the words of their experience, they embodied this energy, thereby acting as a lightening rod or conduit for it to be brought into the earth. They were magicians and alchemists who could transform the archetypal frequencies of energy, which they had absorbed during their journey through the Galactic Center, into matter. They knew the secret of immortality. On most occasions, these alchemists would use the medium of storytelling and poetry recitation and employ geometric shapes in art to influence the soul, conscious of the people and hence their destiny."

As he talks about the power of geometric patterns to inspire consciousness, I find myself picturing the thousands of crop formations that have appeared around the world since the late 1980s, the same period of time during which enhanced access to the Galactic Center has been possible. These complex geometric designs, often seen in the southeast corner of Britain, have been credited to the brilliance of extraterrestrials, the influx of sound waves, the power of the collective unconscious, and, more recently, to circle makers, a group of highly skilled human crop artists.

Whatever the source, these formations are known to carry specific vibrations of energy that steep us in detailed archetypal patterns, influencing us on a profoundly cellular level. If string theorists are correct and consciousness consists of waves of varying frequencies impacting every moment of our lives,[12] then these universal shapes are, like a mold, forming and defining our thoughts and consequently our actions—in other words, determining the actualization of our world.

Although archetypal patterns impact us at the deepest level, the final product is totally unique, expressing the perfection of diversity within

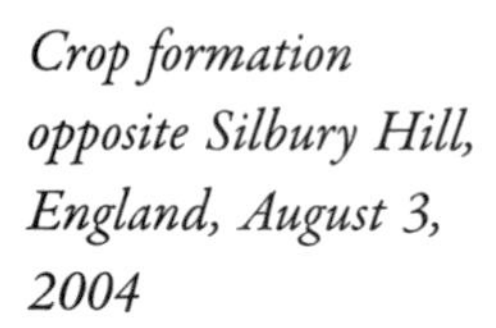

Crop formation opposite Silbury Hill, England, August 3, 2004

the universal plan. I look down at the grains of sand around my feet and marvel at how each is intricately different from the others, honed by the wind and the sun over years, and each is as beautiful as the next.

My dear friend, acknowledging the insights I have just received, continues: "This same Galactic Center has been visited by seers, yogi, and shamans throughout recorded history as they have traveled along streams of projected consciousness. They describe dimensions ungoverned by the time-space paradigm, where there is access to multidimensional and parallel universes and where the source of all life exists. It has been called heaven, nothingness, zero point, and the void, although it is in no way empty but is instead full of potentiality. A more accurate term for it would be the *quantum plenum.*"[13]

My mind returns to a recent lecture I heard on nonlocality and its inherent quality of interconnectedness. The speaker called this nonlocalized place *phase space,* a mathematical term that describes an invisible location consisting of vibrating fields of probability, where every possible past and future is readily available—in other words, the Great Mother. Within this space there is no matter, and everything there exists as pure light or essential consciousness that cannot be measured. All we can do is be present to its uncertainty. As soon as observation, attention, or analysis is applied, the waveform collapses, probability is changed into an actuality, and particles of manifested reality are produced.

What I learned from this stimulating discussion is that it is only from a linear standpoint that there is a clearly defined past, present, and future. In the world of nonlocal reality, everything exists in suspended animation, the now exists with the past and the future as merely different views of our holographic reality brought into manifestation by simply changing the focus of attention or angle of perception. When viewed in this way, it is easy to appreciate the possibility of altering not only the future, but also the past.

Excited by the links my mind was making, I look directly at my friend and say: "As you describe the journey of sages, shamans, and kings to and from the Great Mother, my understanding of the mysteries of immortality is deepening. I now understand that when I am living in the present, I am immersed in eternal possibilities while experiencing the finite wherever I choose to pay attention at that moment in time. If I change my focus, the old focus of my attention transforms from finite back into the ocean of possibilities, and a new wave of probability is converted into manifested reality or form."

Delighted by my understanding, he shifts his position in the sand. "There is a saying which is often misquoted," he explains. "'As above, so below.' To an alchemist the true message is: 'As below, so above; as above, so below.'[14]

"Life is a continuum, represented by the serpent that eats its own tail, the Ouroboros, reminding us of the continual cycles of death and rebirth that occur time and again until there is only now. In terms of immortality, we see the creative source, or ocean of possibilities, and the creation—manifested reality—as equally honored, interchangeable, and essential to each other. In other words, spirit and matter are just different faces of the Divine. Each gives birth continually to the other at the expense of its own existence.

"The ability to shift frequencies between the higher vibration of spirit and the lower vibration of matter is possible because of the existence of a transformer found not only within every human being, but also at the center of the galaxy. This unique piece of equipment is the heart.

The Ouroboros, which reveals the cyclic nature of our existence (Ouroboros, 1478 drawing by Theodoros Pelecanos, in the alchemical tract entitled Synosius*)*

Without it, our very purpose as creative beings would grind to a halt as we are forced either to live a life of continual unrealized potential or to stagnate, paralyzed by the weight of our own creations. Can you now see why limiting thoughts, lack of joy, and denial are major factors in the formation of heart disease? When we are disconnected from the pulse of our heart, we forget that we are immortal magicians and we readily relinquish our wand and our power. It is time to "re-member" who you are; this can occur only through listening to the call of the heart.

"In more modern terms, this miraculous heart could be called a star gate, transporting us naturally within this multidimensional existence. Anybody who has ever been in love knows of the heart's capacity for transformation. Held within the energetic embrace of your lover, time dissolves, space expands, and anything is possible. Everybody you encounter within this blissful existence appears loving, beautiful, and without a flaw. Here, you experience the essential quality of spiritual existence, and it feels wonderful. Now the challenge is to create that heartfelt space wherever you go, resonating with the energy emerging from the heart of the Great Mother, which embraces us all equally; the energy we call love.

"In order to embody such a state of immortal freedom, we must first build and then increasingly occupy our light body or Ka, the vehicle that allows us to escape the confines of time and space."

The mention of the Ka reminds me of Tom Kenyon and his book the *Magdalen Manuscript*[15] in which he describes the alchemical teachings of the Egyptians in their quest for immortality. They understood that each of us have three bodies: the Khat or physical body, the Ka or light body, and the Ba or the celestial soul or higher self. Through deep, inner work and alchemical practices, they aspired to activate each of the body's chakras, thereby sending energy up a central connecting pathway, the Djed, which is positioned along the spine. The process, known as the raising of the Djed, is clearly synonymous with the raising of the dead, or the transformation of base consciousness into the golden consciousness of enlightenment. It is also the reason why the sacred warriors in the *Star Wars* stories are called Jedi knights, implying that they are alchemists who have mastery of their Djed, or sexual fire.

In order to reach immortality, it was taught that there must be an alignment between the energy within the Djed and the Ba, the higher self or spiritual blueprint. When this sacred marriage occurs, there is an influx of energy into the Ka, igniting it to produce a golden raiment or cloak. The immortal body has thus been created.

My desert teacher, conscious of my understanding, takes up the thread. "Such a process toward immortality often takes many lifetimes. Yet at this moment in our history, it is within the reach of all dedicated students of alchemy as long as they remember the warning given by St. Germain, the great alchemist, who said, 'Any developing alchemist must guard against self-delusion and rationalization.'[16]

"There are also specific keys that are essential to the development of the Ka or light body. These include:

- Being present in the now without attachment to the past and future
- Having conscious appreciation of the ability to create through focused attention
- Paying full attention until inspired thought transforms into matter, and then realizing it doesn't matter

- Having compassion and acceptance for our creations
- Having an appreciation that it is through our creations that our consciousness grows

"The masters of the Great Work of alchemy describe the twelve stages of the development of the Ka in terms of cycles of inspiration and expiration. Both the masculine and feminine aspects of our being are involved in this hero's journey, with the feminine providing the force for the journey and the masculine, the focus.

"The phases of inspiration ask that we:

- Listen to our intuition or inner wisdom
- Nurture and support our dreams and ideas

- Own our power, both physical and spiritual
- Accept the dual nature of existence and face the mirrors of our dispossessed self
- Bring any dream into manifestation and celebrate

"These five phases of inspiration involve the development of the ego-hero who journeys until his dreams are fully realized and he becomes king or sovereign of his own creations. But this is only half of the cycle and, on expiration, the ego-king must prepare to die, offering back to the ever-expanding Great Mother, the eternal source, the wisdom of his earthly experiences so that she in turn can give birth to the new sun or ego, ensuring that the cycle is continued.

"Thus the stages of expiration require us to:

- Enhance the process of contemplation and introspection
- Strengthen the power of love to ensure that we turn away from the outer world and back toward our core
- Descend into the boiling cauldron of the Dark Goddess, where the meat of our 'stories' dissolves until only the bones remain
- Meet those aspects of the self that became separated through

shame and fear and accept them into the heart without judgment
- Transform the physical form into spiritual essence
- Complete the development of the magician's wand or Djed, through which we can reach the Ba
- Through the sacred marriage, merge with the Great Mother in the time that is only now."

He waits as the impact of his words fully pervades my awareness. Deep within my heart, I know what he says is true, and yet, at the same time, I am surprised to feel a nagging concern from within, which demands to be heard: "What if I miss the boat? What if I get it wrong?"

He laughs at my honesty. "See how quickly your limiting and unresolved fears override the sense of knowing and excitement? This is normal and helps you to recognize those parts of your consciousness that have become separated over time and that await reintegration. For example, remember an idea or dream that so excited you that you made plans to bring it into form and make it reality. Then remember that when this happened, the manifested dream was less than rosy, attracting emotions that sullied the experience. Despite your original enthusiasm, you withdrew your attention from this creation and moved on to something else.

"Yet in doing this, you abandoned a part of your consciousness, which could be seen as one petal of your flower without which the fullness of the bloom will always be incomplete. It is inherent in you to return and tear away the outer coverings of the 'story' until you can glean the gems of wisdom (consciousness in action) that are waiting within. Unless we do this, we can never gather enough strength to build our light body and hence be the creators of the new world that we are destined to become.

"The Dark Goddess is in residence now, inviting us to enter her fiery cauldron so that we can recover the gems of our experiences from all of our creations over the past twenty-six thousand years. As the light is released, we will understand the teachings about the resurrection of Jesus, who inhabited his Ka after his death, encouraging us to follow his example."

"What happens to the parts of me that I choose to try to ignore?" I ask.

He answers without hesitation. "This is a time to complete karma and extricate yourself from the many illusions you have developed about yourself and the world in general. The subpersonalities or archetypal complexes that you do not possess will possess you, becoming the origin of your creations over and over again until you take notice. This is not a form of punishment, as some would choose to believe, saying, 'This always happens to me,' but instead it is a gift, reinforcing the impetus to remember who you truly are.

"The Dark Goddess or Crone does not have a good reputation; she often trails confusion, death, and destruction in her wake. Yet during this transformative process, her love for us is stronger than ever, for only she knows that in order for a new world to be born, the old must be allowed to die. Even now, her vulturelike nature is tearing away the meat of our old, redundant stories until only the bones of our essential consciousness remain." I can sense her presence physically, and it feels remarkably reassuring.

"So what you are saying is that the more I call home those parts of me that have become separated for whatever reason, and find a way to accept their presence in my life, the more energized will my light body be and the greater the chance to embrace my dreams for the future?"

In an instant I realize there will not be one messiah or guru to lead us. *We* are the ones we have been waiting for! It is our combined consciousness, the very thoughts I am entertaining at this moment, which will shape our destiny. Through this portal of opportunity we are being handed the keys to the doors of heaven so that we can bring cosmic riches down to the earth. Then a thought crosses my mind: What would have been the reaction of the king-shamans or priests of the past if their sacred position in society was challenged by the ability of their subjects to make the same journey and speak directly to the source? It would perhaps beg the question, "Why do we need a king?"

We are truly entering a time of personal empowerment, when self-consciousness and respect for the unique contributions of every individual, symbolizing the forthcoming Aquarian age, is superseding the old Piscean model in which an individual's existence was dictated by the beliefs and dogmas of the few.

It suddenly dawns on me that an extremely clever plan has been set in place to be activated at this time. Each soul has been given a unique part of the jigsaw puzzle which, when brought together with all other parts, will create the fullness of our future. Yet in order for this to happen, we as a species must learn to value and encourage the contributions of every individual; no one piece is more important than another. I can see this concept is a true reflection of the ideals of the fifth world of ether, where all aspects are accepted equally and loved by the Great Mother.

"But why wouldn't everybody want to take advantage of this extraordinary time in our history?" I ask, bemused.

He leans forward again and draws some numbers in the sand:

40 percent   40 percent   15 percent   5 percent

"As we move from the fourth to the fifth world, change is offered through the dissolution and transformation of the old. Yet despite the inherent gift of free will to the human race and the fact that many people feel that their life is a struggle and they wish for change, when the sand is wiped clear and choice is offered, 40 percent will instantly recreate what has just been swept away. Another 40 percent, unaware that they have a choice, will find that the challenge of the void is too much for them and they will instantly fall asleep or become unconscious mentally, remaining in this state until an external source reawakens them. Fifteen percent will become confused, which will be expressed as irritation, frustration, and disorientation, and 5 percent will understand, recognizing the opportunity to be vanguards and light bearers of a new creative cycle, both for themselves and for the world in general."

His words make me realize that we are all offered the chance to enter a black hole, an unknown and seemingly chaotic reality requiring us to be willing to shift tracks, change our frequency, and learn a brand new tune. "But why would we not want change if offered something better?" I ask my patient teacher.

He replies with a sigh: "When you have forgotten the sweet embrace of the immortal self, you will cling to your possessions—material, emotional, or mental—as lifelines, even though these attachments have long since failed to nurture your soul and they appease only the personality. It's so hard to appreciate that these changes aren't happening to you—but are mainly being directed by you, immersed as you are in a mythology that a higher power directs your every move. It is the ego that maintains the separation, the soul that knows better, and the divinity that calls us home."

He looks out as the dying sun's rays set the sky on fire, an omen for the times ahead. "I don't deny that there will be challenges as we begin our journey back to the Galactic Center. There is a large part of us that enjoys its freedom and does not like the idea of being under the control of the cosmic in-breath, or inspiration. Yet, when we listen to our hearts, the truth is clear: It is time to dissolve back into the ocean of possibilities so that we can cast our line afresh and bring a golden age of unity and peace into existence.

"The ancient people knew this time would come and left encrypted messages for us to find. These were hidden within mythological tales; ancient songs; poetry; and sacred sites, especially those employing sacred geometry and artwork. Today, we are seeing a revival of interest in all things sacred and mysterious, with even the scientific world dedicating serious research time and money to the study."

He looks down at the sand that he has recently cleared. "I took great pride in creating my beautiful picture, and yet that same pride should not prevent me from allowing each grain of sand to return to its source. Nothing is permanent. The dissolution of our present state of consciousness has already started and is mirrored by events around the world, both natural and man-made, causing us to release unhealthy attachments to the old way and embrace the spirit of cooperation and compassion.

"Watch the birds and other winged animals—including the dolphins, which are the birds of the ocean—for unlike the four-legged, they are not tied to the earth. They are forming closer and closer ties to the human race, communicating directly with our spirits and urging us to stretch our ethereal wings and rise above any delusion of limitation."

My awareness drifts to thoughts of my beautiful garden, which is commonly visited by hummingbirds, blue jays, and finches, while in the skies above, the red-tailed hawks glide in the warm thermals. My eyes meet those of my companion, and any separation between our souls disappears in waves of joy.

"The journey to the new world is through every heart," he says. "Whether found in the Galactic Center, the nucleus of a cell, the body of a curious and adventurous child, or within the breast of your joyful hummingbird. Remember to celebrate every moment, take nothing for granted, and let your inner wisdom be your guide."

His final words are almost lost as I feel myself being bathed in an overwhelming wave of love, which seems to be simultaneously emanating from and flowing into my heart. Everything seems perfectly clear, and for once, my head isn't directly involved.

"Seek out the knowledge of previous episodes of great change that are often hidden in your mythology," he advises. "To the uninitiated, these cultural tales are merely stories told to children, and yet they were designed poetically by your predecessors in order to awaken your dormant memory when the time is right. Now is that time. The symbols contained within these tales resonate with your eternal mind, awakening you to a truth that is beyond the understanding of the logical mind. While the intellect will look for a list of things to do, the heart will be fed by all the wisdom it needs. Each cultural myth carries archetypal frequencies of consciousness embodied within the lives of that culture's gods and goddesses. When shared, these powerful vibrations stir the smoldering embers of our soul, and we remember.

"At the same time, study the craft of alchemy, the science of the mystics, for these great scientists knew how to turn the base consciousness of ignorance into the golden consciousness of enlightenment and immortality."

He stands slowly, and I know our time together is coming to a close. He adds, "This is the moment you've been waiting for. It's hard for you and others to appreciate the amount of preparation you undertook to be ready for this particular incarnation. You will meet many others who signed up for this great adventure; together you will act as conduits to bring a new frequency of consciousness to this planet."

And with that, the vivid scene disappears. I emerge from a deep meditation and, on opening my eyes, see the Milky Way snaking its way across the night sky high above the Big Island of Hawaii. At that moment, I know that my life has changed forever.

This meeting in the desert took place some years ago. In retrospect, it was obvious that I had few clues as to where to begin my search for archetypal patterns of transformation, but, as has always happened in my life, the areas to explore began to emerge and the pieces started to fall naturally into place.

As I deepened my study of mythology, it became clear that even though the personalities of the main characters of different stories were often altered to accommodate the cultural bias of a particular time, the deeper meaning remained unchanged. Unraveling the mystery allowed me to go behind each story and attempt to extract its essence. It is this light of wisdom that I now offer, with the hope of creating an energetic wormhole into the multidimensional nature of our existence.

1

THE CREATION MYTH

Following my magical evening in the desert, I began to search for mythological stories that could provide me with clues as to how the ancient people dealt with the challenges and chaos that often seemed to accompany times of global transformation. It became clear, as my teacher had suggested, that much of the mystery was embodied in tales surrounding the archetypes of the gods and goddesses and that their trials and triumphs mirrored our own journeys toward spiritual enlightenment.

I was, however, amazed to discover that these immortal beings not only had left clues as to how to survive the turmoil of change, but also actually offered us the opportunity to know the same eternal existence. Their stories contained specific details concerning the extraction of the elixir of life or the ambrosia of the gods, essential for the attainment of immortality.

Yet I also learned that, as for all good secrets, the path was closely guarded and accessible only to those who found their way through purity of being, mastery of their desires, and a detachment from the result. This was underscored by the following advice from the Mayan Popul Vuh: "The Truth is Hidden from the Seeker and Searcher."[1] And it was reinforced by this text from the Gospel of Thomas:

> The Knowledge has been hidden from those who
> wish to enter . . .
> Be as innocent as doves and as wise as serpents[2]

When I first read these words, a resonance reminded me that despite any inept attempts to control my world, everything works out exactly as it must. About ten years before my visitation to the higher realm of the desert I glimpsed in that transcendent meditation, I had become fascinated by all things Mayan. I had traveled the Ruta Maya, studied the Maya calendar, and ensconced myself in the secrets of the crystal skulls, which are commonly associated with these ancient people. It was during my travels that I just happened to meet a man who owned a café in a remote area of Belize. Over numerous cups of coffee, I was enthralled by his stories encapsulating the mysteries surrounding the ancient Maya. Certain descriptions remain in my mind even today:

> The Maya are masters of illusion, keepers of the portals to the holographic universe. In the ancient past, they set time-space locks on these doors so that they could be opened only by those whose level of consciousness ensured that a wise and respectful use was revealed. What archaeologists and the media call a "discovery" is merely the opening of a portal as humankind's awareness reaches a level of consciousness that honors the sacred enigma of the gods.
>
> The Maya also hid their precious artifacts in areas that even today are easily overlooked as insignificant, such as within a piece of uncultivated land. At the same time, they marked portals into the mystical world of the Maya by leaving things unfinished, broken, or out of place. Hence, we see that despite the skills of the master craftsmen who built the magnificent tomb for Lord Pakal, in Palenque, Mexico, one corner of the lid of the sarcophagus is damaged. To the Maya, there are no coincidences and everything has a purpose, offering us a focal point that leads us out of the realms of illusion and into our multidimensional reality.

Today, wherever I travel in the world, I remember these teachings and recount many occasions when I was deeply led into the ancient traditions of a culture by not following the tourist path or by not agreeing with the archaeologist's modern explanation of sacred sites. Rather, I trusted my own intuitive guidance. With this in mind, the exploration of the mysteries that are unraveled in the following pages seeks to open similar portals into the Great Mother so that we may also see beyond the illusion and shadows and know ourselves fully as eternal light beings.

MEETING THE MOTHER

It feels natural to begin our journey in the "place" that exists before a separation into masculine and feminine from which everything is born and will eventually return. Throughout the world, this is described as light, God, the source, or the Great Mother—an all-inclusive energy without differentiation. In today's language, it is described as the holographic universe, the void, or nothingness, which embraces all that has been and ever will be in the essential now.

Unfortunately, such unlimited vastness is often difficult for our small minds to grasp, causing us to seek definition in terms of size and function and leading to our personalized view of God. This propensity to see everything through our lens is the very reason why it is said that the name of God should not be spoken, for in doing so, we limit its existence immediately. Similarly, our tendency to see everything from an objective viewpoint relative to ourselves causes us to externalize this encompassing energy and hence perceive God as separate from ourselves. In truth, we exist within the oneness, affecting and affected by every frequency of life that connects to this eternal source.

As the ancient myths continue, they reveal that the Great Mother gives birth to herself, the Immaculate Conception, and comes to be called the ocean of possibilities, the sea of milk, the Mother, and in modern terms, the quantum plenum or the phase space. To an alchemist, she is the One Thing, which is seen to be synonymous with the soul, imagination, or collective unconsciousness.

Her appearance within world mythology reveals her as Demeter to the Greeks, Lakshmi to the Hindus, Tara to both the Indians and Tibetans, Kwan Yin to Buddhists, Nu Kua to the Chinese, Cybele to the ancient Turks, Diana to the Romans, Freya to the people of northern Europe, Astarte to the ancient people of the Middle East, Isis and Hathor to the Egyptians, Inanna to the Sumerians, and Mary to the Christians, to name just a few.

Of the three aspects of the Triple Goddess, it is probably the Mother who is best known and accepted within most cultures; she is seen as the nurturer and a provider for others who gives from her abundant ocean of possibilities. Commonly represented as a large-breasted woman, she was epitomized, as we have mentioned, by the goddess Diana (the Greek Artemis), with her whole torso covered in breasts. Her name translates as Di-Anna (Dinah), or "grandmother of God." Two thousand years ago, Diana had an extensive following throughout the lands bordering the Mediterranean; a magnificent temple was built in her honor at Ephesus in what is now Turkey.

As Christianity spread across this area, however, these sacred places of worship were either destroyed or transformed into churches dedicated to Mother Mary. Di-Anna was now demoted to Anna, the grandmother of Jesus. Yet the worship of the goddess Diana lives on in the Festival of Candles, which takes place annually in her honor on August 15 throughout Europe and symbolizes her promise of eternal life for those who follow her loving example. It is interesting to note that this same date is now associated with the Christian church's celebration of the Assumption of Mary, in which the message is essentially the same.

In other cultures, we find the Mother is represented by a multinippled and horned animal such as a cow. This representation gives rise to a popular emblem of the Mother aspect called the horn of plenty or cornucopia, from which pours the fruits of the earth.

In Hinduism, the Mother aspect of Kali is symbolized as a white-horned, milk-giving moon cow, while in Egypt, Hathor, the Mother of all gods and goddesses, is commonly depicted wearing cow's horns and offering her breasts with both hands. The high esteem in which Hathor

Hathor, the heavenly cow

was held in the Egyptian culture comes from her representation as a heavenly cow whose udder produces the Milky Way and who daily gives birth to the sun god Horus-Ra, her golden calf.

Thus it is no surprise to learn that as the Israelites left the land where the goddess Hathor was honored as the source of continual sustenance, they took the opportunity to melt down the gold and build a golden calf while Moses was away receiving the Ten Commandments.

This same theme of the Mother being depicted as a milk-giver gave rise to the name Italy, meaning "calf land," suggesting that it had been birthed by the Goddess. Even earlier myths from Japan, the Middle East, and India speak of the universe being curdled into shape from cow's milk, leading to the Hindu creative myth Samudra Manthan (the churning of the sea of milk), which we will discuss later.[3]

One final example of the Mother energy is the Greek goddess Demeter (the Roman goddess Ceres), who was goddess of agriculture and of the harvest, sister of Zeus, and mother of Persephone. Her name comes from the Greek letter *delta,* which means a "triangle," and *meter,* which means "mother." Together they represent the yoni, or vulva, of the Great Mother, ensuring that she has an important position among

Demeter, goddess of agriculture and the harvest

the deities of the Greek culture. As daughter of Cronus and Rhea (also a Mother goddess), she taught humans how to sow and plough fields, thus bringing to an end their nomadic existence and enabling them to create the first planned societies.

Demeter's symbols are a multiseeded head of wheat; the cornucopia; and a golden, double-headed ax. Her symbolic animals include the pig, dragon, serpent, and turtledove, all of which are associated with other powerful Goddess figures. As we will hear later, she is best known for the grief she experiences after her daughter is abducted into the underworld and for the creation of the Eleusinian mysteries, which teach of the mystical steps that can be taken to achieve immortality.

Despite a sometimes delusional belief that the Mother is a benign being whose bountiful gifts are simply there for our pleasure, it is important for us to understand that she represents the force to create. Every droplet of her ocean is in constant, dynamic readiness, continuously interacting with every other droplet of creativity. It is she who provides the energy behind every manifested form we see in the world, and it is her energy that causes a blade of grass to quiver and hot lava to flow from a mighty volcano. All our thoughts and actions are motivated by her powerful expression known as emotions, which, as we all know, can change in an instant, transforming any outcome dramatically. This is the face of the Hindu's Shakti, whose presence as the Mother is chaotic and powerful. It is she who embodies this force to create and without her, even the most focused attention on our dreams and aspirations will amount to nothing.

The Mother demands a continuous exchange of energy with all those who dare to throw their fishing line of desire into her waters, warning them that once they eat of the fruit of good and evil[4] and know the duality of life, there is no turning back. From that point on, we are bound to cycles of life that bring into form the Mother's unmanifested seeds and then return the essence of that form back to the Mother's eternally changing ocean.

Thus we see the emergence of the other two faces of the Triple Goddess, one representing her creative powers as the Virgin and one her destructive energy as the Crone. Along with the Mother, they create the feminine Trinity.

Through her different aspects, the Mother maintains her fruitfulness,

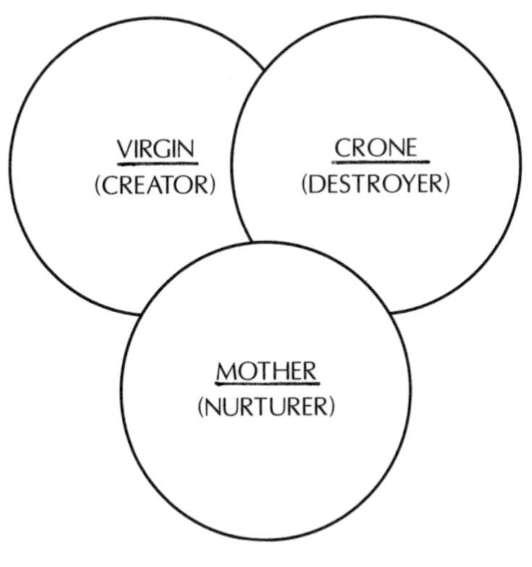

Triple Goddess

feeding us with abundant opportunities for bringing spirit into matter and then nurturing herself with the gems of wisdom that emerge from our experiences.

The Triple Goddess

Traditional wisdom tells us that the Triple Goddess was honored as far back as 25,000 BCE. It has only been in the past two thousand years, under patriarchal rule, that the Goddess and everything associated with the feminine (intuition, emotions, and rhythmic cycles) was denigrated. These qualities, however, are essential for our soul's journey, whatever our gender.

Intuition maintains the connection to our soul's blueprint.
Emotions act as the force behind all creativity and transformation.
Cycles prevent stagnation and ensure continuous growth through opportunities for death and rebirth.

Even a superficial study of the last millennium reveals the extent of this inequality. Of an estimated fifty thousand to one hundred thousand people killed between 1400 and 1700 CE in the European witch hunts, most of these victims were women.[5] The list, however, also included men who professed a spirituality that did not align with the dogma of the Catholic Church at that time. These persecutions arose from a number of factors, many of which were irrational and fear-driven but which fueled the fire that women and their mysteries were the cause of disharmony, whether at home, in the community, or even around the world.

During the past fifty years, tremendous strides have been taken to redress these imbalances. Through the spotlight of the media and the Internet, the important role of women within society is increasingly being observed, with many inconsistencies slowly but surely being addressed. These changes have a direct relationship to the manner in which men and women alike embrace their own inner feminine nature, an issue that is a daunting prospect for some.

The Triple Goddess in Mythology

Representations of this Triple Goddess are found throughout traditional texts, in which they appear as Parvati-Durga-Uma (Kali) in Hindu tradition, Ana-Babd-Macha (the Morrigan) in Ireland, and Hebe-Hera-Hecate in Greece. All three aspects of the Triple Goddess are found as Guinevere in the Arthurian Legend, as the druidic Diana Triformis, and as the Fates to the Romans.

In the ancient city of Glastonbury, England, the image of the Triple Goddess is revealed naturally in the landscape:[6] The gently rising Wearyall Hill is seen as one of her outstretched legs; the Chalice Well, her pregnant belly; and the famous Tor, her left breast. This shimmering grassy hill, visible for miles around, contains the markings of a three-dimensional Cretan labyrinth, a sure sign that the Tor was considered a sacred portal into the underworld, the home of the Crone.

Beneath the Goddess's womb lie the ruins of Glastonbury Abbey. With its straight lines and towers, its design contrasts starkly to the curves and flow of the surrounding landscape and reflects the patriarchal influence present in the thirteenth century during its construction. Yet beneath the main structure of the abbey is a more ancient ruin called the Mary or Lady Chapel, which is dedicated to the Virgin and built around the principle of circles, to embody the perfect proportions of a *vesica piscis.* This beautiful chapel honors the yoni, or vulva, of the Goddess, symbolizing the portal from which we are born and through which we will eventually return to complete the cycle of creativity and fulfillment.

The final representation of the goddess in Glastonbury is marked by a small hill called St. Bride's Mound found at Beckery, west of the main town. This simple rise in the landscape is seen to symbolize the head of the baby as it appears from between the legs of the Virgin. Metaphysically, it also represents the head of the fully evolved hero-king who, as lover, willingly dives back into the vagina of the Crone to complete his journey.

Despite the fact that this land is privately owned, is not acknowledged

Glastonbury Tor

as a sacred site, and is located behind a disused sewage works, it was once the gateway to the mystical Isle of Avalon. This ancient location, symbolic of rituals and celebrations involving death and rebirth, emanates an ethereal energy where the veil between the worlds is thin and there is easy access to the other dimensions. In the past, Avalon was indeed an island, with waters lapping its green pastures gently. St. Bride's Mound, situated on the riverbank, symbolized the completion of a physical pilgrimage and

The mists of Avalon

the beginning of a spiritual one. Here, travelers would be prepared by the maidens of St. Bride to enter their own underworld before crossing the water by boat to meet their destiny.

The inconsistent preservation of the sacred sites in Glastonbury is typical of many such locations around the world, some faring better than others. Matters, however, are changing as people of both genders recognize the need to restore the sacred balance between the masculine and feminine and hence find spiritual union within themselves.

THE BIRTH OF THE ONE MIND: THE MASCULINE

Out of the Mother aspect of the Triple Goddess, the One Thing, the ancient creation myths tell of the birth of her consort or masculine counterpart, known to alchemists as the One Mind. This represents the focus to create, the source of attention, direction, and discernment—the conscious mind without which the feminine force to create stays in its amorphous form.

Together, the feminine *force* to create and this *focus* give rise to *intention:*

FORCE + FOCUS = INTENTION

A useful analogy is that when we decide to cook a meal, the ingredients, including the heated oven, embody the feminine *force* or energy, while it is the masculine *focus* that decides on the recipe and eventually brings the meal to the table. Working together, they create our concept of reality.

When focus and reality align we call it *synchronicity* and recognize it as part of the journey toward the experience of oneness, especially when the outcome is positive. Yet when the same process of combining force and focus results in a perceived negative situation, it is not uncommon to hear a defiant denial of any connection between the thinker and his or her manifestation; most prefer to see the outcome as mere coincidence.

Within the realms of creativity, we cannot pick and choose what we decide to own. Everything that causes us to have an emotional reaction reflects a part of our own spiritual blueprint. This is based on the understanding that our purpose on earth is to express fully into form our unique aspect of the Great Mother so that she may know herself through us. Hence, as we move through the world, we project our energetic blueprint into the environment. Naturally, this attracts people and situations that will bring the blueprint into manifestation. This Law of Attraction is not derived from will-based intention but from the undeniable intention of the soul, causing us to receive what we need, not what we want.

When our emotions and senses are activated by something pleasant, we recognize the part of the self that is mirrored in the situation and we experience positive resonance. Often, however, we are less willing to be accountable for our creations when the form that manifests causes us to feel angry, defensive, irritated, and fearful. In this case, there is a great tendency to deny the connection and judge the offending situation or person. Yet whatever we judge in others is what we fear to find in ourselves.

Like it or not, whatever belongs to us will not go away, for it is awaiting integration, and at this specific time, twenty-six thousand years of our creations are lining up to be remembered. Indeed, the road to immortality asks that we meet and embrace all aspects of ourselves until we can stand fully radiant in our light body, until the physical form no longer hides its brilliance.

In summary, there are two primary archetypal energies, called different names, out of which all creation is born and from which our dreams will ultimately reach manifestation.

One Thing + One Mind
Divine Mother/Goddess + Divine Father/God
Soul + Spirit
Chaos + Structure
Emotion + Logic
Force to Create + Focus to Create
Power + Purpose

In the world of duality both aspects have equal importance and each impacts everything we do. The masculine face provides our world with structure and stability mainly through rules, laws, and beliefs, while the feminine face provides movement and opportunity largely through creativity, inspiration, and intuition. Problems arise when one aspect is given preference over the other, causing the one that is suppressed to rise up in order to reestablish its rightful position.

When the world becomes too structured and lacks spontaneity and growth, the feminine face offers turmoil, which shakes things up and offers new perspectives from the realms of imagination. In a similar way, when mayhem or confusion is rampant, events occur that force us to make decisions or set priorities, bringing masculine order out of the chaos.

RIGHT RELATIONSHIP

A continual flow of energy between the masculine and feminine principle is the force that creates healthy, reciprocally rewarding relationships where all aspects are respected and honored.

The Iroquois Constitution

Many ancient traditions used their understanding of these diverse feminine and masculine forces to create peaceful, just, and prosperous societies. They selected women as representatives of the Mother Goddess to bring forth creative inspiration and intuitive wisdom, while the men were chosen as rulers to turn these visions into reality through their practical strength. Even the Iroquois Constitution, which became the basis for the Constitution of the United States of America,[7] declared that the chiefs would be chosen by the women of the tribe, for it was believed that only women could selflessly decide what was best for the people. We can only imagine the transformation that would occur around the world today if women and only women were allowed to vote!

If we step more deeply into the story behind the inception of the Iroquois Constitution and its Great Law of Peace, Kaianeraserakowa, we meet Jikonsahseh, a woman who, despite her neutrality, sells provisions to the warring factions. It is not until the "peace maker" Deganawidah shows that her trade is in fact supporting the killing that she decides to sell the message of peace, rather than one of war. Known as the "mother of the nations," she is accorded the position of being one of the original clan mothers charged with the decision to both elect and remove the chiefs. Deganawidah believed that it is only women who know the hearts of men, who are connected to the earth's abundance, and who know the pain of burying those they love, including the children they have suckled at the breast.

According to this constitution, only women have the wisdom to decide whether any battle or war is worthy of the cost of life. This truth has been reinforced several times around the world recently, where peace has emerged from women coming together across the divide, driven by an inherent love that has no restrictions and where all colors, creeds, and religions are accepted as part of the Great Mother.

As Jean Reddemann, a wise Native American teacher says, "It is prophesized that a thousand years of peace will come when women heal their hearts."

The Ego Is Born

As the creation myth continues, we learn that the One Mind and the One Thing express their love by breathing life into each other, thereby creating a vesica piscis, out of which is born the *ego,* the creative light of manifestation.

This vesica piscis, a sacred geometric shape, provides a doorway into new states of awareness because the ego, a divine child born through the power of love, is destined to change the consciousness of its parents, for it contains both aspects of its parents: masculine spirit and feminine soul. The former maintains focus and direction while the latter offers all the energized ingredients to allow the goal to be realized.

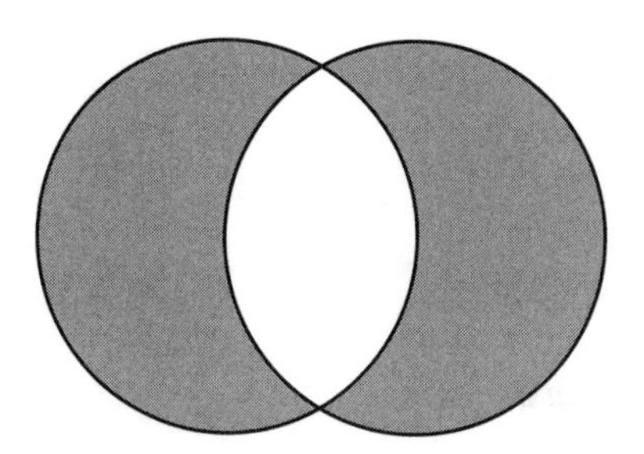

Vesica piscis

Without an ego there can be no spiritual journey, for:

It is through the growth and development of the ego
that energy is transformed into matter, creating what we
perceive as reality.
It is through the ego's willingness to die
that the essence of the experience is gifted back to the source,
reminding us that nothing really matters!

This cycle between spirit and matter or heaven and earth is also known as the hero myth in which the ego passes through several stages of evolution and dissolution, taking on different roles as it dives in and out of the Great Mother's ocean of possibilities. The intimate balance between the masculine (focus to create) and the feminine (force to create) and their relationship to the signs of the zodiac can be expressed as follows:

- **Virgin:** the seed of potential, our creative blueprint **(Pisces)**
- **Puer:** the innocent boy-child, inexperienced but eager to learn **(Aries)**
- **Mother:** nurturing and support for the journey **(Taurus)**
- **Hero:** grasping the dual nature of consciousness, physical and spiritual **(Gemini)**
- **Crone:** meeting and embracing the mirrors of our existence **(Cancer)**
- **King:** the crowning and full expression of our blueprint **(Leo)**
- **Virgin/Triple Goddess:** introspection and contemplation **(Virgo)**
- **Lover:** compassion, fairness, reason, and self-discipline **(Libra)**
- **Crone/Triple Goddess:** embracing inner riches **(Scorpio)**
- **Sage:** wisdom, truth, and insight, based on detachment from form **(Sagittarius)**
- **Triple Goddess:** transforming matter into pure essence **(Capricorn)**

- **Magician:** wand holder, shapeshifter, shaman, living in both worlds **(Aquarius)**
- **Great Mother:** Nothingness, the end and the beginning **(Pisces)**

Sacred Numbers of Creativity

The emergence of everything out of the Great Mother—to which all will eventually return—is expressed in a sequence called Fibonacci numbers, believed to underpin the success of any creative enterprise. Each number, except the first, is produced from the addition of the two previous numbers:

1 1 2 3 5 8 13 21 34 55 89 . . .

The relationship between the numbers is eventually seen to result in the golden mean, in which a number is 0.618 times greater than the previous number, stretching into infinity. This is also known as the Golden Ratio. Such a fundamental mathematical equation for creative evolution is expressed in natural phenomena such as the pattern of seeds in a sunflower head. These seeds are arranged in logarithmic spirals flowing out from the center in both directions. If the seeds along the clockwise and counterclockwise spirals are counted, it is found that the totals represent two successive numbers in the Fibonacci series, such as 34 and 55 seeds. This golden spiral is seen in other natural forms such as the scaly plates of a pineapple, the seeds on a pine cone, and the position of branches around a tree trunk, to mention just a few. It is believed that nature appreciates that the golden spiral is not only pleasing to the eye, but offers quantity without sacrificing quality.

When we review our present understanding of the creation myth, we see that it also follows this ancient system:

- In the Beginning is the Great Mother: **1**
- She gives birth to herself, the One Thing: **1**
- The One Thing gives birth to the One Mind: **2**
- Together, they create the ego: **3**

- Out of the ego come the five stages of manifestation (Aries to Leo): **5**
- These are followed by the stages of destruction into transformation (Aries to Scorpio): **8**
- Together, these stages represent the complete cycle of creativity through death and rebirth (Aries to Aries): **13**

These last three numbers (5, 8, 13) represent the different levels of initiation through which we must pass if we are to achieve the ultimate goal of immortality. They strongly relate to the awakening of not just the seven well-known chakras, but also, ultimately, to the twelve chakras, some of which radiate beyond the physical form. Thus 5 and the pentagram represent the change that occurs when we master and synthesize the four elements so that spirit can be lifted from the confines of a purely material existence. (Astrologically, this is Aries to Leo.)

The number 8, the octagon, represents the number of notes in an octave and symbolizes the integration of spirit and matter, in-breath (inspiration) and out-breath (expiration), leading to the extraction of the consciousness of light. (Astrologically, this is Aries to Scorpio.)

The number 13, the number of transformation, represents completion and death of the 12 with the rebirth of the 1 as the serpent eats its own tail—the Ouroboros. Once we understand the esoteric nature of these cycles, it is not difficult to imagine why 13 has always been depicted as unlucky by those who desire to wield power over others. Yet when we acknowledge 13 as being auspicious, we remember that first and foremost we are immortal, without the need of an outside authority to make us whole. (Astrologically, this is Aries to Aries.)

As the journey unfolds, it becomes clear that our own creative process follows the exact proportions of sacred geometry and mathematics and that despite our frequently limited and subjective perspective on life, everything is perfect.

The Essential Nature of Duality

Before we delve more deeply into the hero's journey, we must learn to appreciate the importance of cycles in our lives and how they reveal the intricate dance between two opposite poles of existence and the duality in which we live. Increasingly, there are those who wish to eliminate duality in order to embrace the unified field, without wondering why such a construct was created in the first place.

Let us remember that the new world of ether represents a fusion of polarities where both are equally accepted and necessary, and it is through the sacred marriage of opposites that we generate the light required to achieve an eternal life. Those who advocate union without first embracing their own shadow are not only practicing spiritual bypass, but also they can ultimately cause destruction to both themselves and others as their disconnected parts seek attention and acceptance within the heart.

It is through our ability to master and embrace these opposing forces that spiritual growth occurs, where each aspect seeks not only its own fulfillment but also that of its partner, through a continuous process of give and take. This exquisite interplay of energies is reflected monthly by the cycles of the moon, and is etched into the messages, in myth and legend, left to us by our ancestors.

2
RHYTHMS OF THE MOON

Despite the "one giant leap for mankind" taken in 1969, there is still much we do not know or understand about the moon's mysteries. Often represented as having a feminine nature, you may be surprised to learn that a study of world history reveals that both moon gods and moon goddesses were often worshipped simultaneously. Indeed, in many early cultures it was the moon god (not the goddess) who conferred fertility and nurturance, with women and farmers praying to this deity for fruitful abundance.

It was only with the transition from lunar to solar worship approximately three thousand years ago that there was a distinct emergence of a moon goddess. The priests of the day, however, were eager to distance themselves from anything feminine, and thus she was portrayed as a fearful goddess of death, which ignored her beauty and her natural rhythms. Thus we meet the Greek goddess Artemis, who is depicted as a dispassionate huntress carrying her bow (the crescent moon) and accompanied by her faithful hunting dogs. This same stereotype is repeated within Mexican mythology, in which the moon goddess is seen as a demon roaming the skies at night, seeking victims to devour. Indeed, many funerary traditions even today include the practice of adorning those who have died with crescent-shaped amulets, in hopes of ensuring favor from the moon goddess as the journey through the process of death begins.

More recently there has been a tendency to see the moon's influence mainly through the lens of the watery emotions. Yet it can be said that the earth's constant companion is indeed encoding us with much more profound messages that encourage us to appreciate the cycles of growth and transformation; to honor death and rebirth; to develop healthy, interdependent relationships; and to remember our eternal nature.

MOON CYCLES

There can be few people who have never been impressed by the changing faces of the moon in the night sky, even in the most light-polluted environments. Yet outside traditional cultures, it is seldom taught that each moon phase represents a different phase of the hero's journey toward spiritual enlightenment.

Each month, with the help of the moon's energy, we are encouraged to plant seeds of inspiration, celebrate our successes, harvest and absorb the fruits of our endeavors, and allow the old to die eventually so that the new can be born. If we can plan our lives to correspond to the unique features of each phase of the moon, we will find that life will flow with far greater ease.

Moon phases

The phases of the moon are described in the following ways:

NEW MOON PHASE: UP TO 3½ DAYS AFTER THE NEW MOON

Impulse: to germinate and emerge

Archetype: puer (as noted earlier, the innocent boy-child who is inexperienced but eager to learn)

Action: This is a time to plant new seeds of inspiration. Even though the final manifestation of these seeds may be unclear, there will be an instinctual urge to encourage these ideas so that they can see the light of day.

CRESCENT MOON PHASE: 3½–7 DAYS AFTER THE NEW MOON

Impulse: to move forward through focus

Archetype: puer to hero

Action: As the desire to emerge into the light continues, there is often an opposing force within that seeks security and the comfort of the known. Through focused attention we come to recognize that any struggles, whether internal or external, are merely the means of strengthening our spiritual resolve.

FIRST QUARTER PHASE: 7–10½ DAYS AFTER THE NEW MOON

Impulse: to build and decide

Archetype: hero

Action: Now there is a strong determination to define our goals and seek self-individuation, causing old patterns of fear, denial, and limitation to rise up and be acknowledged so that important choices can be made for the future.

GIBBOUS MOON PHASE: 10½–14 DAYS AFTER THE NEW MOON

Impulse: to improve and perfect

Archetype: hero to king

Action: This is a time for self-reflection and evaluation that allows us to refine and produce the perfect bloom. It is inevitable that self-doubt and self-criticism will also arise, but with objectivity this should only enhance the end result.

FULL MOON PHASE: 15–18½ DAYS AFTER THE NEW MOON

Impulse: to seek integration consciously and achieve fulfillment

Archetype: king

Action: This is a moment for celebration, allowing all to see the full bloom of success. Yet the festivities may be delayed if we continue to seek the perfect relationship, result, or effect without realizing that everything exists perfectly within. The full moon allows us all to be seen in our true light.

DISSEMINATING GIBBOUS MOON PHASE: 3½–7 DAYS AFTER THE FULL MOON

Impulse: to distribute and convey

Archetype: lover

Action: Now we are given the opportunity to share our experiences with others, allowing what was relatively external to become embodied in reality. Now we must walk our talk and realize that there is a fundamental difference between success and fulfillment.

LAST OR THIRD QUARTER MOON PHASE: 7–10½ DAYS AFTER THE FULL MOON

Impulse: to revise and reevaluate

Archetype: sage

Action: This is one of the most difficult phases: We are asked to ferment in the deeper and darker energies that surround our creations. Shame, fears, and old beliefs come to the surface to be honored and assessed along with redundant masks and veils. This is a time to absorb the essence of our experiences and let go of the story. This is not a good time to start something new.

BALSAMIC MOON PHASE: 10½ DAYS AFTER THE FULL MOON AND UNTIL THE NEW MOON

Impulse: to distill and transform

Archetype: magician

Action: As the final phase of this cycle comes to its close, the essence of experience is offered back to the Great Mother and it is time to rest. This phase is often an inward process associated with mystical experiences, healing, and transformation.

Now we enter three days of darkness, when the illuminated face of the moon is hidden from the earth and the moon is conjunct the sun. Then one day, something begins to stir again within the dark waters of the Great Mother and a new moon or child is born through the Virgin. A new cycle begins.

Each of us was born within a particular phase of the moon, specified as the relationship between our natal sun and moon. This can be located within each person's astrological birth chart and reveals the particular emphasis of the spiritual journey in life. (See chapter 12 to calculate your lunar phase.)

RECIPROCALLY REWARDING RELATIONSHIPS

One of the most powerful messages that the moon offers every night is the dance played out between the darkness and light. We are often so focused on the ever-increasing light that we fail to acknowledge that, at the same time, the darkness is dying. Eventually, at full moon all we see is light, and it could be said that the dark sacrificed itself so that the light can grow. Then, within hours, a small sliver of darkness appears, and over time, the light is consumed by the darkness until it is seen no more.

This theme of sacrifice in the name of love is played out in many myths in which the nurturing mother dies as her hero-son leaves home to find his fortune. The second part of the story is rarely told, however: the son returns as king and is willing to feed his mother with the seeds of his endeavors, dying slowly to his past so that eventually she can give birth to his son and heir.

In the final decades of this fourth world, we have seen ourselves drawn into the darkness of the vagina of the Great Mother as our outer light diminishes, and we feed ourselves with the fruits and flesh of our own experiences so that we can eventually offer our essence to the source. This is the true nature of any reciprocally rewarding relationship in which both parties gain through willingness to accept the principles of death and rebirth. This is a very different concept, for in this scenario, both aspects of the cycle are willing to give so that the other can flourish.

This view is seen in the stories surrounding the moon tree. This sacred tree is often portrayed as being protected by two winged animals, the lion, which is the symbol of the interplay between dark and light, and the unicorn, which is the symbol of unity and immortality. Together, these animals reflect the paradox of divine creation: the dynamic flow and tension between the two opposite poles of existence (the lion), nurtures the central, eternal energy (the unicorn), which in turn is the original source of the duality. In mythological images, the branches of the tree are depicted as either being full of fruit or having been removed, leaving only the trunk visible—depending on the phase of the moon that is represented at the time.

This interplay of energies symbolized by the moon tree is spoken of in hymns to Ishtar, the great Babylonian moon goddess, daughter of the moon god Sin. She is known as the sacred tree, and she carries Tammuz, her consort and son. He symbolizes the new, green growth from which the precious fruit appears, representing the continuity of life. Their close interdependence is captured in the deep understanding that she needs him to fertilize her by acting as her spouse, and in turn, she will give birth to him in the form of her son.

This description of a healthy, symbiotic relationship reminds us of the interplay of energies that must exist within us for the production of the elixir of life. It shows that when two opposite poles of existence, such as our own masculine and feminine natures, work together in perfect synergy and harmony, the result is a self-perpetuating flow of energy that has no end. As we will see later, this is represented by the toroidal energy of the heart.

It is interesting to note that in many moon-based cultures, the felling of a tree is an important ritual signifying the willing sacrifice of the dying god to the Great Goddess. Indeed, for the Egyptian god Osiris, as for other gods, a truncated tree was the resting place of his coffin, representing both the mother lovingly embracing her son and the mandatory demise of the son to receive such a blessing. This tradition of cutting down a tree is still performed by those who celebrate Christmas. In the homes of these believers, the tree is decked with lights and baubles to signify both the death of the old king and the welcome of the new sun or son.

ETERNAL LIFE

Who would have thought that the unassuming moon would have such a strong connection to the secret of immortality? Many traditions are rich with such stories.

Take, for instance, the tale of Chandra, a Hindu moon god:

> Dressed in white and carrying a crescent moon–shaped bow, Chandra is seen as the guardian of soma, the nectar of immortality. This precious elixir is fermented from the milky white juice of a plant that grows exclusively upon his sacred mountains.

This association between moon gods and mountains is a common theme and reminds us of the close relationship between the energy within the body's chakra system that runs along the spine and the energy that is thought to pulse through the core of many great mountain masses. Could it be that mountains, like our bodies, act as lightning rods, drawing consciousness between the upper and lower worlds? In addition, perhaps there are optimal times for this transmission—times associated with the cyclical dance between the moon and the sun.

As we will see later, the ancient people believed this to be true, leading them to build towers, pyramids, and ziggurats, which functioned as transformers, converting spirit into matter and matter into spirit.

The next link between the moon and the elixir of immortality comes from China, where moon cakes are still served at a festival that falls on the fifteenth day of the eighth moon of the Chinese year. This date equates to the September equinox and to what is known as the Harvest Moon, a time of abundance and prosperity.

Chang'e is a beautiful female immortal who comes to earth to help her people. Here the account varies depending upon the version. All of them agree that Chang'e eventually swallows a pill or magic potion containing the elixir of life and finds herself floating up to the moon, where she takes up her position as goddess.

Her companion on the moon is a character who appears in many of the lunar legends, especially those originating in Asia, Central America, and the Australian continent. While those who live in Europe and North America see the dark troughs on the moon's surface as the "man in the moon," individuals in other parts of the world are convinced that the troughs in fact form the figure of a rabbit. Chang'e's assistant is known as the jade rabbit and this is his story:

> Three fairy sages transform themselves into pitiful old men and beg for something to eat from a fox, a monkey, and a rabbit. The fox and the monkey both give food to the old men, but the rabbit, being empty-handed, jumps into the blazing fire, offering his own flesh so that the men may eat. The sages are so touched by the rabbit's sacrifice that they let him live in the moon palace, where he is given the honored position of grinding the jade elixir of immortality.

This simple story carries an important message: Through the sacrifice of our flesh (our earthly attachments) in the fiery cauldron of the Great Mother, we receive the promise of eternal life.

THE ELIXIR OF LIFE

What is this potent potion that is so highly prized throughout all traditions? It has many names, which include amrita (Indian), the fountain of life (Christian), the elixir of immortality, dancing water, ambrosia (Greek), pool of nectar, and the quintessence of life.

Alchemists throughout the ages have aspired to produce this elixir through chemical processes that combined the four elements of earth, air, fire, and water into the fifth element of ether, which, interestingly, is associated with the new world of the fifth sun (as we learned in the introduction). Many have believed that when these white drops (liquid gold) are drunk, they offer good health, eternal youth, and immortality to those who consume them.

Others believe that this magical potion can be produced naturally by a woman during the act of sex. It is known as the divine nectar of the sacred waters and is believed to contain the fountain of youth. Researchers have shown that this fluid, ejaculated from glands in the anterior vaginal walls, contains an enzyme that extends the life of cells and hence has been called the enzyme of immortality. This same nectar is even mentioned in fairy tales in which the princess is "awakened" by the kiss of her handsome prince—a metaphor for an intimate, sexual embrace.

Represented by the lily or the lotus, the elixir of life is linked strongly to the sexual, Dark Goddess Lilith. Like other Crones, Lilith was instrumental in teaching her priestesses how to produce this fountain of youth through sacred sexual practices as a means of conferring immortality on those who came into their presence. It is interesting to note that Lilith was the first wife of Adam, but she refused to stay and share her juices with him when he denied her equality within the marriage bed. Surely, this was an opportunity for immortality that was lost in a single moment!

In Hindu tradition it is believed that amrita is produced from the pineal gland during deep states of meditation, which involves the raising of the kundalini through the chakra system. Teachings profess that just one drop of this potent elixir is enough to conquer death. This is exactly the same process described by the Egyptian alchemists. They saw the raising of the Djed through the activation of the chakras as the means by which enough energy was generated to create a dynamic vessel that drew in the Ba, or celestial body, waiting above the crown chakra. Through the sacred marriage of these two powerful forces, the Ka was ignited and immortality was achieved.

Many believe that every time we complete a cycle of creativity from birth to death, which can happen several times in one life, we stimulate the pineal gland to produce a small amount of amrita. This, in turn, helps to develop and strengthen our Ka, or light body, which leads to our nurturance coming less from the denser frequencies of physical food and increasingly from our ingestion of the higher frequencies of light emitted by plants, animals, the celestial bodies, and ether itself.

As the Ka grows, we find ourselves less affected by the challenges of aging; we look younger and stay healthier. Over time, we are able to live a more multidimensional existence in which time is a less influential factor in our lives. Eventually, we see time for what it is: a Temporarily Induced Mind Experiment (TIME).

MYTHOLOGICAL ASTRONOMY

So why is the moon traditionally connected to such an important component of our spiritual transformation? What can we learn from its mysteries that will enable us to extract this precious elixir? It is clear that many of the cyclical activities of the moon influence our psyche deeply, even though outwardly we may poke fun at those who allude to an association between human behavior and this celestial body—for who hasn't been awestruck by the vision of a deep orange ball rising above the horizon at sunset or by the first sighting of the crescent moon in the dark sky? Unlike its solar equivalent, this heavenly mass appears to have no obvious purpose and yet is inherently revered by all.

It is well known that the moon's silvery sheen is due purely to the reflected light of the sun. Subsequently, it offers a pure and perfect mirror for anything projected upon it and yet has no attachment to the result once the projection is withdrawn. Keeping this in mind, it is easy to see how the moon could and has been used as a focus for all our projected psychological shadows. Just as the Galactic Center, the ocean of possibilities, offers us whatever beliefs we choose to take to it, so the moon reflects whatever is in our deep unconscious. Some individuals sense mystery, others a joyful familiarly, and a few the face of madness.

This deep, intimate relationship to the moon is expressed in words such as *lunacy,* from the Latin *luna,* and *mental,* from *mens,* both meaning "the moon." Historically, such terms represented an ecstatic union with the spiritual essence of this heavenly body but have been distorted to reflect a disease rather than a state of bliss. The same root gives us the words *menses, menstrual,* and *menopause* because of the strong link between the cycles of creativity within a women and the cyclic nature of the moon. It is clear, however, that in our modern vernacular, the link more commonly made is between a woman's cycle and her mental mood!

Even in the past fifty years, the moon has played a major role in shifting the consciousness of humankind: When we first saw the earth rise over the horizon of the moon during the Apollo voyages, our perspective of this beautiful planet's place within the solar system changed and somehow we knew we were no longer alone. At the same time, many of us fell in love with this fragile, blue globe, which heightened our awareness of the honored position we hold as guardians of the moon, a fact we seem to keep forgetting.

Let us now explore some of the moon's mysteries from an astronomical viewpoint.

Does the Moon Have a Dark Side?

Many of us have heard that there is a "dark side" of the moon, although it would be more accurate to say it has a hidden side. We all know that the earth travels around the sun and the moon travels around the earth. Yet it is almost incredible that the rotations of the earth and moon on their axes are in such perfect symmetry that they always show us exactly the same face of the moon; the opposite face is hidden permanently from us here on earth. It was not until 1959 that the Luna 3 probe exposed the secrets of the moon's hidden face when it sent back pictures of the moon's mountainous terrain.

Could this hidden side of the moon reflect the "hidden" side of us, which can been seen only with our inner eyes and which enjoys the sun's full embrace in the darkness before the new moon?

Solar and Lunar Eclipses

All ancient astronomers knew the importance of an eclipse in world history. Indeed, the Mayan calendar accurately calculates the date of every solar and lunar event from 3114 BCE until the solstice on December 21, 2012.

An eclipse exists due to an alignment of the sun, moon, and earth. A solar eclipse always occurs at the time of a new moon, when the moon is located directly between the sun and the earth. A lunar eclipse, on the other hand, accompanies a full moon, with the mass of the earth standing between the sun and the moon, preventing most of the sun's direct rays from reaching the moon's surface. Solar and lunar eclipses always accompany each other fourteen days apart, although which appears first is dependent upon the orbital patterns of the sun and moon at a specific time.

During a lunar eclipse the moon often appears blood red, which caused many in the past to fear the demon that unleashed such horror. They did not understand that the hue is due to the shorter wavelengths of the sun's light passing through earth's atmosphere and coloring the moon's surface. The astronomical events were deemed so important that fearful Roman emperors tried to deflect the personal effects of an eclipse by murdering their top-ranking statesmen on the advice of astrologers. Other devious leaders used their prior knowledge of a solar eclipse to trick the gullible populace into believing that they could control the sun's movements.

Of the two types of eclipses, a lunar eclipse has always been more commonly associated with malevolence and mystery. Many cultures believe that during an eclipse, the moon is swallowed by a mythological creature. To the Maya, the jaguar was the culprit; to the Chinese, the aggressor was a three-legged toad. For many others, the culprit was a dragon. Whatever the beast accused of the crime, an eclipse brought with it a profusion of bad tidings. Indeed, it was not uncommon for citizens to run into the streets shouting, screaming, and banging drums to frighten away the evil spirits.

To the ancient Maya, the jaguar ruled the night; its spotted pelt representing the starry sky.[1] During a lunar eclipse, it was believed that the jaguar's mouth was wide open, enticing the people into the underworld to face those parts of themselves still within the darkness, and hence glimpse the potential of eternal life. This same belief is held by modern-day psychology in which, during a lunar eclipse, the glow of the full moon—our outer expression in the world—disappears, leaving us to face aspects of our psyche that live in the shadows. This causes dismay to some and ecstasy to others.

In Hindu tradition, Rahu, a power-seeking demon depicted as having the head of a dragon and the body of a snake, swallows the sun or the moon, causing an eclipse. According to legend, during Samudra Manthan (the churning of the ocean of milk), Rahu, disguised as a god, sits between the sun and moon and manages to drink some of the amrita (the drink of immortality). Before the divine nectar can pass his throat, however, Mohini (the female avatar of Vishnu), cuts off his head, causing his body to remain untamed and unruly (symbolized by the serpent). The head remains immortal, and it is believed that this part of Rahu swallows the moon, causing an eclipse. The eclipse comes to an end when the moon passes from the opening of his neck. Traditionally, the head of this great demon is called Rahu and the tail is known as Ketu.

This myth offers us a further glimpse into the deeper mystery of an eclipse. It suggests that during such an event, we are offered the chance to partake of the drink of immortality, but only after we have mastered our erratic serpentine energies, a practice deeply ingrained in the alchemical process of self-realization.

The North and South Nodes of the Moon

This Hindu story is a powerful metaphor for what actually happens during an eclipse. Rahu and Ketu, the dragon's head and tail, symbolize astronomical points in the sky called, respectively, the north and south lunar nodes.

We might assume that every time the moon passes behind the earth, a lunar eclipse occurs. Yet the moon's orbit around the earth does not exactly follow the earth's orbit around the sun on what is known as the ecliptic plane. The moon's passage can vary by five degrees above or below the ecliptic, intersecting this invisible orbital line twice a month: once when it is descending (Ketu), and once when it is ascending (Rahu).

Approximately four times a year, the alignment between the sun, moon, and earth occurs close enough to these nodal points for us to experience either a lunar or solar event, although not all of these result in a total eclipse.

In astrology, these placements have been given great importance. The south node, shaped like a cauldron, represents past karma and what is being left behind. The north node embodies the skills we are developing,

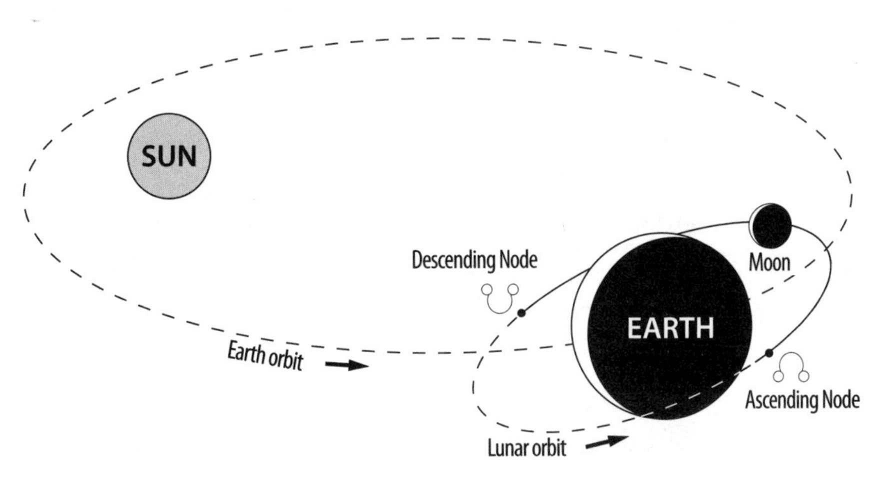

The north and south nodes of the moon

reminding us of why we are here. (See chapter 12 for guidelines to the meaning of your own natal nodes.)

The Celtic Great Year: The Saros Cycle

The moon is affected not only by the earth's gravitational field but also by the powerful force of the sun, which causes it to "wobble" on its orbit around the earth. This wobble causes the nodes to move backward or retrograde through the astrological signs.

It takes approximately 18.6 years—nineteen eclipse years—for the moon's nodes to regress one complete revolution through all the signs. This cycle was considered extremely important in Celtic and Mayan mythology because it acknowledged the deep relationship between the cycles of the moon and an eternal source of creativity. In ancient times in a town called Kildare in Ireland, there was a perpetual fire dedicated to the virginal goddess of the hearth, Brigid. It was watched over continuously by nineteen priestesses representing the nineteen-year cycle of the Celtic great year, or the Saros cycle. Each took their turn staying by the fire for nineteen days, knowing that on the twentieth day the fire would be tended by Brigid herself.

The true meaning of the archetype of the Virgin is to be complete unto yourself without the need for another to make you whole. Numerologically, 19 is the number associated with this profound description of the Virgin. The number 1 represents beginnings and 9 represents completion. In other words, 19 is seen to represent the cycles of death and rebirth, symbolized by the snake that eats its own tail, the Ouroborus.

It was also believed by Brigid's followers that the physical fire absorbed the exquisite energy generated during the loving relationship between the sun and the moon and transformed it into a continual source of creativity and abundance that was available to everybody upon this earth.

The fire was extinguished during the sixteenth century, but it was relit in 1993 and has remained alight since then. This fire in County

Kildare, Ireland, is a constant reminder that unless we acknowledge and honor the powerful influence of the celestial bodies upon our lives, the flow of creativity and abundance available to all of us will eventually dry up. As we will see later, a similar fire was attended by the vestal virgins of the Roman Empire until it too was extinguished by those who did not understand the importance of the perpetual fire. (We all know what happened to the Roman Empire!)

As mentioned, according to the Saros cycle, eclipses will occur within the same sign of the zodiac every 18.6 years. The eclipses do not appear at the same geographical location, but appear 120 degrees farther around the globe. It is only after fifty-six years that what is known as a triple Saros occurs—that is, the eclipse is seen both within the same sign and in the same location.

From archaeology, we can determine that the builders of ancient sacred sites knew about this cycle. When the first phase of Stonehenge was erected during the latter part of the fourth millennium BCE, the builders dug fifty-six holes that housed moveable wooden posts just inside the outer earth bank surrounding the site. Today, the posts are no longer present, although white disks cover the holes, which are named Aubrey holes after John Aubrey, who discovered them in the seventeenth century.

The posts are known to have been markers for the calculation of lunar eclipses, and thus they reinforce the suggestion that such an event is not merely a spectacle, but also energetically influences the consciousness of the earth and its people. Scientists are only now able to measure these so-called subtle energies but are finding that in fact there is nothing subtle about their effect.

Moon Standstills

Another important feature of the movement of the moon is that there are times when the moon appears to be at a standstill, an event that occurs when the moon is approximately at right angles to its nodes.

During the winter months, the full moon appears ever higher in the sky until it reaches its maximum height above the visible horizon. This is the full moon that is closest to the winter solstice. In contrast, during the summer months, the full moon appears ever lower in the sky until it reaches its lowest position. This is the full moon that is closest to the summer solstice. This movement is opposite that of the sun: The sun reaches its highest point above the horizon at the summer solstice and its lowest point at the winter solstice.

Due to the Saros cycle, however, every 18.6 years the moon reaches a maximum and minimum point above the horizon. These points were perceived by ancient peoples as being high and low points of fertility for both the land and its people. Hence, in 2005 and 2006 we experienced a major lunar standstill season: the full moon at the winter solstice was the highest it had been in the sky in nineteen years and the full moon at the summer solstice was the lowest. In 2015, approximately 9.5 years later, we will experience a minor lunar standstill season when the reverse occurs. The full moon at the winter solstice will be at its lowest point in the sky and the full moon at the summer solstice will be at its highest.

The ancient people believed that the dance between the sun and the moon was very important due to the powerful gifts each body brought to the land and its people. Hence, they recorded both lunar and solar standstills (at the solstices) during the construction of their sacred sites. These sites include Stonehenge, the Callanish Stones on the Isle of Lewis on the west coast of Scotland, and Chaco Canyon in New Mexico.

Dark Nights of the Moon

Nearly all traditions agree that the process of death and rebirth requires three days, symbolic of the three dark nights before the moon returns to the sky. It is therefore inexplicable that the Christian festival of Easter, which represents the death and resurrection of Jesus, is in fact positioned to take place at the time of a full moon and not a new moon. We can only imagine how the effectiveness of the Easter message has been diminished by this distortion. The cover-up is more complete by the representation of the resurrection (the miraculous ignition of Jesus's Ka) through yellow bunnies and chocolate eggs. The greatest fear that many organized religions have is that we will discover the importance of death, the underworld, and hell in the process of our spiritual transformation.

Calendars: Measures of Time

By now, it has become obvious that the moon's predominant role in many ancient cultures is deeply embedded in both mythology and science—so much so that it is often difficult to separate the two. This is also true of the measurement of the passing of days.

Until the establishment of the Julian calendar during Roman rule, most calendars were based on the lunar cycle of 355 days, with the 356th day starting the thirteenth month. In 1582, the Gregorian calendar replaced its Julian counterpart, firmly assuring that the international method of recording dates would be based on a solar and not a lunar cycle.

Yet several ancient peoples maintained their link to the old ways by continuing a *lunisolar* calendar. These groups included the Jews, followers of the Hindu faith, and various cultures in ancient China. Islam is the *only* culture that still uses a purely lunar calendar. Indeed, the crescent moon is an important component of the Islamic faith. This should not be a surprise, for during the time of Abraham, father of both Judeo-Christianity and Islam, the people worshipped the moon god Sin. They followed his guidance and appreciated the importance of the moon's cycles to the prosperity of the people.

Today there are many calls for a global return to a lunar-based calendar, realigning us to the phased nature of the moon. At this time of the fusion of polarities, however, there is a place for honoring the importance of both the sun and moon and for following the doctrine of the fifth world of ether.

MYTHOLOGICAL ASTRONOMY AND SACRED SITES

To understand the influence of an eclipse and other planetary movements upon consciousness, it is valuable to appreciate the specific features that accompany the building of many of the megalithic sites around the world, including those found in Britain, Ireland, Central and South America, Egypt, and throughout Europe.

- Each was built to receive and transform waves of archetypal and high-frequency energies that passed into the site, bringing new frequencies of consciousness to humanity.
- The materials used in the construction have strong piezoelectric qualities: They are able to collect, store, transform, and emit any energy in the vicinity.
- Each was built with a profound understanding of the energy grid system or aura surrounding our planet and embodied within it. Hence, many of the sites are situated on the intersections of major ley lines, the portals to the grid system of the earth.
- Most sacred sites, including modern-day crop formations, are located on soil that is composed mainly of limestone or a similar porous substance. This allows archetypal energies to be transferred into the water in the soil (in a homeopathic form) and to be absorbed eventually by every person around the planet through the flow of the mighty oceans.
- Just as any electronic device is constructed to manage varying frequencies of energy, so these formations were designed specifically through the use of sacred geometry, ensuring a maximum result with minimal detrimental side effects.
- Each was constructed in reference to the movements of not only the sun and moon, but also other planets and star systems such as Venus and the Pleiades.
- The effectiveness of each formation is heightened by the presence of all the elements. Hence, the correct proportions of earth, water,

fire, and air were provided either by the builders or by the natural surroundings of the site.

- Most of the crop formations appearing around the world are designed in exactly the same way, so that they function in the same manner as the stone megalithic sites.
- The energy or emotion brought into a sacred site is enhanced and absorbed by the surrounding energy fields. If we bring fear, that is what will be enhanced; if we sing and dance, then joy is transferred. We always have a choice.
- Despite historical findings, many of these sites were built before humankind took physical form in order to transmit universal information or consciousness into Gaia, our earth, a living breathing being.
- Many were held within the higher dimensions of the holographic field, unseen and "undiscovered" by the human race until our consciousness reached a vibration high enough to draw them down into the third dimension.
- Most sacred sites were built at a time when stones could still "walk." Hence, there was no need to create fancy harnesses or large teams of people to move them; all that was required of humans was knowledge of how to communicate intentions to the consciousness of the stone people. They then cooperated. This gift of communication can still occur through the heart when we have no other agenda except to be in harmony with all inhabitants of this earth.
- Every site acts as a lightening rod, radio receiver, or communication tower among the dimensions, using frequencies outside our normal sensory range but certainly perceivable by those with heightened sensory skills, such as children, animals, and birds.
- In its most pristine form, the specifics of the site act as a filter for received information, which passes into all the levels of the earth's grid: etheric, astral, mental, soul, spiritual, and universal.
- The most important receiver is the soul grid or dodecahedron,

which connects us to the heart of the Great Mother and the field of unity and harmony. Individually, such a connection enables us to remember the true nature of our existence as spiritual beings.

- There are a number of interested parties who prefer that we remain ignorant of both our ancestry and the communications from other planes of existence, especially those from the star people or extraterrestrials. These interested parties often discredit the importance of sacred sites to our evolution.
- The sacred sites, whether man-made such as Stonehenge or natural such as Uluru, call on us to remember who we are and to bring our joy and laughter to the area, permeating the grid with light.

With all of this in mind, it is clear that significant events such as a solstice, equinox, or eclipse open a specific portal or doorway into the multidimensional nature of the universe. When this happens, archetypal frequencies of energy are downloaded to our planet from the celestial sphere, offering us the chance to be bathed in new and enriching waves of consciousness, which are essential for the time ahead.

FOLLOWING THE NATURAL RHYTHMS

Some of the oldest references to moon worship come from ancient Sumer, which was established in an area south of modern Baghdad around 5000 BCE. Here, the moon god Nanna, symbolized by a crescent moon, was seen as the torch of the night, renewing himself constantly like a snake and illuminating primeval darkness.

Nanna taught the people that there was a perfect time for everything and that by flowing with these cycles, life became easier. Thus he ruled rhythms and cycles; tidal waters; human emotions; the fertility of crops; the endocrine system; and, of course, the menstrual cycle, which so closely follows the cosmic cycle of creativity. Without these cycles, the ancient people knew that their land and lives would become dry and barren. This is a fact that we must remember as we attempt to control menopause, menses, sleep cycles, and the timing of birth and death.

Following Nanna's guidance, most individuals of the time lived by these rhythms. They saw the waxing or crescent moon as a most propitious time to fertilize crops, ideas, and people. The word *crescent* comes from the Latin *creare* meaning "to create." In those times, it was during a waning moon that harvesting and pruning took place, allowing the old to die and be laid to rest. Among the traditional people, nobody would think of planting crops or starting a new business during this time. Those who lived close to the tidal waters had their own ideas, believing that a good birth was one connected to an incoming tide, whereas to die "well" was to expire as the tide went out.

It would be interesting to project how health and productivity are transformed if we adopt these rhythms, both in our personal lives and in the arenas of commerce, medicine, and education.

The Lightning Rod

During the time of Nanna, the people started to build towers or observatories called ziggurats. It is believed that these were designed both as a place to study celestial movements and to worship the moon from the temple that sat atop the structure. These towers continued to be built right up until 2100 BCE, when the famous Tower of Babel was constructed in what is now southern Iraq. The Babylonians had continued the Sumerian tradition of lunar worship through their moon god Sin. It would be wrong, however, to assume that these were unsophisticated people baying at the moon, for records show that the Babylonians were a highly developed civilization with a deep understanding of the sciences, including medicine, chemistry, alchemy, botany, zoology, mathematics, and astronomy.

The decision of the Babylonians to build the Tower of Babel with mathematical precision could be hypothetically associated with a profound understanding of universal energies, alchemical processes, and a desire to experience the unified field of nonlocal reality—in other words, to know the immortality of the gods.

The Tower of Babel

This, then, offers an interesting twist to the story told in Genesis 11:6–8. From the text, we are told that God is concerned when he sees that the people have built a tower with its top in the heavens so that they can remain together as one unit. He says:

> "Truly, they are one people with the same language. This is the beginning of what they will do. Hereafter they will not be restrained from [anything] which they are determined to do. Let us go down and confuse their language so that they will not understand one another's speech. So the Lord scattered them from that place all over the Earth and they stopped building the city."

From this story comes the origin of the verb *to babble,* interpreted as "to speak incoherently." Many understand this text to mean that God saw that in having one voice, humankind could challenge his will and that they would self-destruct without his guidance. Hence, as one interpretation tells it, he stepped in for our own good, causing separation and the formation of diverse nations around the globe in order to create peace.

If this interpretation is correct, it is clear that that the plan didn't work! Perhaps the greatest cause of humanity's destructive nature is the very fact that we do not have a common language to unite us. This is not merely a matter of verbal expression. What is missing is our ability to *hear* the common language of our souls.

Yet we might ask these questions:

- What if, during this ancient time, there was not just one divine being but many lesser gods and goddesses who came to earth some four hundred thousand years ago as extraterrestrials to influence the development of the human race? Zecharia Sitchin and others call these beings the Nephilim, Nefilim, Elohim, and the Annunaki, meaning "those who from heaven to earth came" or "the shining ones." The place of origin of these beings is seen as Nibiru or Marduk, meaning the "planet of the crossing."[2] When the story is seen in this light, it is perfectly possible that it was one of these godlike beings who was not pleased with the fact that the populace had found a way to access the vast abundance of the Great Mother, through their understanding of the complex movements of celestial bodies such as the moon, and chose to destroy their pathway to enlightenment.
- What if the *oneness of language* that concerned God was, in fact, the unified realm of ether in which, as is predicted, we will all be able to "hear" each others' feelings and thoughts through the purity of our hearts?
- What if the language that was confused by the angry God was

not the language of words, but the language of DNA, the means by which we download the information of consciousness?

To support this hypothesis, there are several factors that we now know about DNA: Only 10 percent of the genome codes our so-called genetic make-up. The purpose of the other 90 percent is still relatively unknown; indeed, it is called "junk" DNA. Yet it is also believed that junk DNA plays an important role in our nonphysical communication, a storyline that only now is starting to be explored.

At present, each chromosome consists of a double strand of DNA. Yet according to Barbara Marciniak's book *Bringers of the Dawn*,[3] there was a time when the genetic information, the building blocks for transforming spirit into matter, existed within twelve strands of DNA. Then, due to genetic engineering by "reptilian gods" (which, some say was twelve thousand years ago), the system was rewired into a dual pattern, causing the appearance of junk DNA. At the same time, the rewiring caused us to forget that we are truly creators of our perception of reality and that we are in fact as powerful and immortal as any who see themselves as gods.

We know that DNA has its own energy field and acts as a miniature black hole, drawing information into it from other subtle energy fields. Any disturbance to this central core of energy, including the ingestion of genetically modified foods and any other genetic manipulation, diminishes our ability to decipher the information pouring into this planet at this crucial time.

Could those who built the Tower of Babel be the source of the true derivation of the word *sin,* which gives a clear warning to the followers of the moon god Sin of what will happen if anybody dares to step beyond the authority of those who would prefer that humans remain slaves?

Manna from Heaven

The historian Laurence Gardner offers a further piece of information essential to our appreciation of the importance of these towers or ziggurats. He states that in ancient times, the word *temple* meant not a place of worship but a workshop or factory.[4] Hence, he suggests that a temple to the moon god placed on top of a mountain or tower was probably the site of the production of what is now known as *monoatomic gold.* This extremely light substance is produced by the arcing effect that is created when a thermal arc is transmitted across a piece of common gold. Monoatomic gold has been found to be connected to antigravitation teleportation and, when ingested, eternal youthfulness. Was the position of the temple on top of the mountain designed perfectly to receive solar and lunar energies, which could also produce this arcing effect?

It is now believed that when monoatomic gold is absorbed, it stimulates the release of natural hormones in the body, especially from the pineal and pituitary glands, which, as we have mentioned earlier, ensure that there is an enhanced quality and quantity of life. In other words, it causes the release of the elixir of life.[5]

According to Laurence Gardner, it was this same gold, produced alchemically within the temples during certain auspicious astronomical events, that became the substance known as manna from heaven.[6] There is certainly a strong suggestion of this idea when we see that both *manas* and *mana* are derivatives of the word for "moon." It is also fascinating to appreciate that prior to the appearance of manna in approximately 1960 BCE, kings and other leaders ingested menstrual (*men* = "moon") blood in the belief that it was infused with the moon's power of creativity and hence, the elixir of life. We now know that during menses or the moon time, a woman's body produces hormones such as DMT (dimethyltryptamine),[7] which are found in abundance in the menstrual blood. Such chemicals are believed to cause an expanded state of awareness, an experience that is prized by kings and leaders who wish to maintain their spiritual power.

Thus, every month, due to the release of these hormones, a woman is capable of traveling along the path of the priest-kings of old and entering the heart of the Great Mother. There she receives messages and insights that will benefit not only her spiritual evolution but also that of her family and tribe. On her return from this altered state, she shares her visions with the men, investing them with the authority to bring into manifestation the ideas she has received.

In traditional settings a woman in her moon time is considered extremely powerful, for she is capable of purifying not only the physical body but also the "negativity" of the family as she passes through her monthly death and rebirth cycle. For these reasons, the monthly cycle is honored among indigenous peoples. At the other extreme are more modern women who, all over the world, are taught not to respect their monthly cycle in the same way. Through the media they are advised to take a daily pill that suppresses this natural and essential process.

No wonder there are so many women with premenstrual tension (PMS), problematic moon times, and postnatal depression. Our modern lives do not allow the space to integrate these powerful energies, leading inevitably to distress and dispiritedness. It is only when women decide to respect and honor their gifts that the necessary changes will take place in the human race to bring natural rhythm back into our lives.

MOSES AND THE EMERALD TABLET

One final connection between the lunar and solar cycles and the practice of alchemy is made when we learn that Mount Sinai was named after and dedicated to the moon god Sin. As a note of clarification, the actual location of this famous mountain is now believed to be Jabal al-Lawz in Saudi Arabia[8] rather than the Sinai Peninsula of Egypt. This plateau was the location for the festival and celebration of the full moon called Sappatu, which is the origin of the Hebrew Sabbath.

The Bible states that Moses climbed Mount Sinai to communicate to God and hence receive the Ten Commandments for his people (Exodus 24:12). If some theorists are correct,[9] Moses and the pharaoh Akhenaten were one and the same, believing in the One God as the source of all existence. It is known that Akhenaten was a master alchemist and in fact the reincarnation of Thoth,[10] the author of the Emerald

Tablet. This alchemical treatise reveals to the apt student the steps that should be taken in order to turn base ignorance into the golden consciousness of illumination.

Thus we may ask, when Moses-Akhenaten first climbed Mount Sinai and was told by God (Exodus 24:12): "Come up to me into the mount and be there: and I will give you tables of stone," was he, in fact, receiving the Emerald Tablet?

As the story continues, we learn from the Bible (Exodus 32:19) that when Moses returns to the people, he finds that they have created a golden calf, a reference to their previous worship of the Egyptian cow mother Hathor. In his anger, he breaks the original tablets and returns to the mountain so that God can write a new set of commandments. Were these new commandments different from the first set due to the fact that God had decided that humans were not ready to control their own lives?

There is a twist at the end of this tale. According to the Bible (Exodus 32:20), Moses is told by God to cast the golden calf into a fierce fire to produce powdered gold and then give it to the people to drink. Could it be that Moses created monoatomic gold (also known as manna and the alchemist's philosopher's stone) and, by giving it to the people, thereby altered their consciousness?

There is no doubt that this event reinforced the transference from lunar to solar worship, a phase that has lasted almost thirty-five hundred years. Perhaps now, though, we are ready for another paradigm shift that will result in a resurgence of interest in following the ways of Nanna, the Sumerian moon god. We will remember that there is a perfect time for everything and that by flowing with these cycles, life will naturally become simpler.

3
THE HERO'S JOURNEY

Every culture is eager to share the legends of its heroes and their grand accomplishments, knowing that the examples they set act as powerful role models for others to emulate. The archetype of the hero is exemplified in characters such as Ajax, Hercules, Odysseus, Perseus, and Jason, all of whom arise from the Greek culture. From other traditions we encounter the Celtic Cuchulainn, the Egyptian gods Osiris and Horus, the Scandinavian Beowulf, and the English hero King Arthur. In more recent years, stories of heroes have filled our screens. *Star Wars* regales us with the exploits of Luke Skywalker and his sister, Princess Leah, who together epitomize the idea of twin champions. And now we must include J. K. Rowling's Harry Potter, who symbolizes the mystical heroes of old.

All heroes have one thing in common: they are destined to make a journey that is both a physical and spiritual quest for fulfillment. The greatest study of these hero myth pathways was carried out by the late Joseph Campbell, whose book *The Hero with a Thousand Faces* has sold millions of copies around the world.[1]

Within his teachings he reminds us that despite the tendency to place idolized individuals on a pedestal, the hero exists within everyone. It is the part of us that steps out courageously into the unknown, enjoys adventure, overcomes obstacles, and strives toward contentment, whether the focus is on our inner or outer life. During my working life as an intuitive, I have also noticed that heroic acts do not always follow stereotypical patterns, especially when viewed from the soul's perspective. Courage comes in many forms, and the spiritual purpose behind our actions may lie hidden from conscious awareness. Hence, I have personally observed:

- The black sheep of the family owning the darkness so that others may experience the light.
- A strict and even aggressive parent who, on a soul level, encourages his or her offspring to stand up and be strong in their own right.
- The absent parent who chooses not to adversely effect the decisions of his or her children.
- The child dying of cancer whose triumphant smile tells me that their sole purpose was to bring together a family at war.
- A daughter dying of cancer whose illness motivates her parents to examine their relationship, not only with each other, but also with life.
- A mentally ill sibling who possesses the chaos, in order to bring stability to the rest of the family.

These examples may surprise those who prefer to view their heroes through rose-colored glasses. In reality, however, many of us will be surprised by who greets us when we pass over from this life and who truly has offered us unconditional love during our passage on this earth. Think back through your own life and ask, "Who was my greatest teacher?" Can you imagine the courage it takes to be the one who is hated and resented or plays the part judged to be docile and ineffectual so that another soul can grow? It is only when we step back and take an objective look at major events in our lives that we can acknowledge those who made a difference to us.

As we follow the stages of the hero's journey, we are shown the pathway—taken by ascended masters such as Jesus, Buddha, and Muhammad—toward self-realization and enlightenment. Even though it is often described as a single event, this journey consists of thousands of intertwining roads, each offering a marginally different view of our holographic universe. As we travel, we see that the stages express both the masculine and feminine aspects of the hero, reminding us that it is only through the interplay between the force and focus of creation that we will truly bring heaven onto earth and earth into heaven and thus know the oneness of existence.

SAMUDRA MANTHAN: THE CHURNING OF THE SEA OF MILK

This wonderful Hindu hero myth is one of the most famous episodes in the Puranas and is celebrated every twelve years in a festival called Kumbha Mela.

> Constant warring occurs between two groups, the Devas and the Asuras. One group has divine qualities and the other consists of demons. Tired of battling, the Devas call on Vishnu for a solution and are advised to churn the sea of milk and hence produce the nectar of immortality, providing them with the obvious upper hand. There is one snag to the solution, however: They cannot do this without the help of the Asuras, so the two groups strike a temporary truce.
>
> Each agrees to hold one end of the snake Vasuki, which is wrapped around the mountain Mandara, and to alternately pull, causing the milk to churn. The plan proceeds well until all of a sudden, the mountain begins to sink into the sea, and the Devas once again cry out to Vishnu for help. He sends himself as Kurma the turtle to support the mountain from beneath, and the process continues.
>
> The first substance to emerge is a highly venomous poison, a by-product of the pain the serpent experiences when pulled in both directions. The only being who can safely swallow the venom is the god Shiva, who clears the way for other products to emerge from the milk. The Devas and Asuras watch as a variety of gift-bearing gods and goddesses step out of the sea. The final god is the divine doctor who carries the nectar of immortality, or amrita.
>
> As the Asuras surge forward to partake of the elixir, the Devas call

upon Vishnu to distract their enemy while they drink. He agrees and transforms himself into the glamorous and alluring Mohini, drawing the attention of the Asuras away from the cup. Yet Rahu, an Asura, is not fooled by the ploy, and seizing the cup, he swallows some of the drink. In an instant, Mohini turns and immediately cuts off his head, causing only the top half of Rahu to become immortal, while the rest remains untamed and uncouth.

Hence, the Devas achieve their immortality and we receive an important insight from a Hindu perspective of the steps to be taken to reach a state of immortality through right relationship with the Mother and her ocean of possibilities or, in this case, her sea of milk.

The Symbolism of Samudra Manthan

The story clearly illustrates the multifaceted, dual nature of our existence, which must be honored and respected if we are ever to know oneness.

- The sea of milk or ocean of possibilities is the One Thing, collective consciousness, the plenum of potentiality, and a reflection of our multidimensional existence.
- Vishnu represents the One Mind and is seen as the preserver or the all-pervading one who directs and sustains activities within the universe.
- The Devas and Asuras epitomize the duality of life emerging from the union between the One Thing and the One Mind. These two poles of existence, characterized by the positive and negative aspects of our personality, must work together in a harmonic and integrated fashion in order for us to attain self-realization.
- Vasuki, the serpent, symbolizes the force or power inherent in every facet of the One Thing activated by the focusing ability of the One Mind. The story emphasizes that power on its own is neutral and can be used equally by both the dark and the light.

Indeed, it is only when we harness the power of both sides working together that we are able to produce the elixir of immortality. Hear me when I say that "both light and dark are required to work in harmony to bring forth the riches of the eternal world."

- Mandhara, the mountain, symbolizes focused presence, concentration, or paying attention. The word *mandhara* is made up of two words, *mana* (mind) and *dhara* (a single line), which together mean "holding the mind in one line."
- Kurma, the turtle, is called upon to support the mountain (focused attention) when it starts to sink and represents the withdrawal of the senses or focus into ourselves (the inner psyche) just as a tortoise withdraws its head into its shell. This reminds us that without the strengthening qualities of introspection and contemplation (combining focus and force), concentration is unstable and the focus of attention will fall back eventually into the sea of pure illusion.
- The poison reminds us that as we turn within and face our own demons for the first time, it is not uncommon for deep inner turmoil and painful disharmony to emerge. Psychologically, these demons represent suppressed, archetypal complexes that are held within our psyche from other lives or from ancestral karma and have become separated from us through fear and shame. As part of the remembering, these aspects of ourselves must be met, accepted, and integrated before we can progress toward spiritual enlightenment.
- Lord Shiva, who, like the Crone, is often called the destroyer, is also the one who purifies everyone with his name. Symbolically, this is the part of us that is willing to eat away the flesh of our own creations, even the so-called poisonous aspects, in order to absorb what is good and to release the rest.
- The various precious objects that emerge from the ocean during the churning represent the psychic or spiritual powers (*siddhis*) that we gain as we progress spiritually from one stage to the next. The fact that the gods and goddesses are present to distribute

gifts suggest that they know how easy it is for the seeker to be hampered or awed by these powers and the desire to possess them rather than to use them wisely for the good of all concerned. These divine beings help us maintain our focus by giving us only what we are capable of handling at any one time.

- Mohini, the alluring temptress, symbolizes the greatest challenges to the spiritual disciple: pride and self-magnification, which arise from a nonintegrated ego. It is only when we accept that pride maintains separation that we bow our heads in humility and dive back into the unified field of the sea of milk and achieve the self-realization we seek.
- The amrita, or the elixir of immortality, is our ultimate achievement.
- The fact that it is only Rahu's head that becomes immortal speaks to the Hindu understanding of reincarnation. This says that the essential wisdom of our experiences will merge with the eternal sea of milk, while the untamed and poorly integrated parts of ourselves will return to earthly existence for further refinement.

This beautiful description of the path toward spiritual enlightenment or eternal life highlights the challenges and opportunities that we all face at this particular time. Residing within the fifth and final phase of the fourth world, we are all guided to complete karma and embody the essence from all of our creations over the past twenty-six thousand years so that we may move beyond our separation, even just for a moment, and taste the elixir of immortality.

This traditional Hindu tale teaches that the ability to exist within the "now" is accessed only through the dynamic tension created when the forces of attraction and repulsion are brought together. In other words, this magical space in which everything and nothing exists is accessed only by the integration of duality and not merely by choosing to live in one pole while ignoring the other.

The now of immortality exists when we remember we were never separate except by limiting beliefs that exist within our minds.

Pat's Story

This important message was reinforced at the funeral of a dear friend a few years ago. As the service proceeded, I watched with fascination as Pat's spirit danced around her coffin, commenting on its quality and proudly introducing me to her natural birth family, whom she had met only after she passed from this earth plane. On several occasions, I had to stop myself from laughing out loud as Pat exuded the joy of a soul free of the restrictions of the physical body.

She even moved among the mourners, offering them words of encouragement and love, urging them not to grieve her passing but instead remember their happy times together. Then, as the Catholic priest began his sermon, weaving traditional doctrines with personal sentiments, she settled under the pulpit, concentrating hard on what he was saying, nodding occasionally—until he started to talk about sin and shame.

"Our dear sister Pat's sins have been forgiven, her shame has been washed clean, and she has been allowed to enter God's mansion," said the priest.

Immediately, Pat's spirit became very animated, and in a voice so loud I was sure all the church could hear, she exclaimed: "No, no, that's not the truth! The only shame is that I wasn't taught that I was always welcome in God's house! I now know it was only my beliefs sustained through fear that caused me to feel unworthy and to maintain the separation for so long."

Turning to me, she looked directly into my eyes and said, "Never let the illusionary state of fear get in the way of your true, eternal connection to the source. There is nothing you need to do or change to be accepted in the kingdom of heaven. This is not a place but a state of being in which your luminous self already resides, waiting for you to step over the threshold of uncertainty, and remember."

4

THE CELESTIAL DESIGN

The stage is now set to undertake a spiritual journey that is essential for this particular time in our history. It is time to feed your heart with all the gems of experience and wisdom from the past twenty-six thousand years in order to charge your Ka with the energy of light. It is time to retrieve small parts of your awareness that have been the magnet for the creation of your dreams and ideas and yet have been abandoned when things haven't worked out quite the way you planned.

Despite your expectations, your consciousness has been transformed by the unfulfilled event. Therefore, only by getting to the heart of the matter or the core of the story is it possible for you to extract the pearl that waits within and offer it up to the Great Mother as your humble gift to the collective consciousness.

In truth, this cycle of turning our dreams into form and then back into the pure essence of consciousness often remains incomplete. Many ideas flounder before reaching the full bloom of success, while at other times we cling desperately to manifested dreams without realizing that our soul's fulfillment is based not merely on material success but also on dissolving form back into pure consciousness.

Yet there is one part of our nature that never forgets: the higher self, the bearer of our spiritual blueprint. It is our constant companion, urging us on when we get stuck; encouraging us never to compromise when it comes to spiritual abundance; and marinating us in compassion, whatever choices we make. Like other aspects of our being, its contribution to our spiritual journey is highlighted in mythological stories of gods and goddesses, reminding us first and foremost that we are immortal beings.

The symbolism of our spiritual quest has been encrypted within many ancient divination systems, including tarot and astrology. Our ability to appreciate the wisdom and knowledge held within these sacred systems is directly proportional to the openness of our heart's mind, which inherently recognizes the universal patterns flowing through all creative processes.

ASTROLOGICAL MYTHOLOGY

The word *zodiac* is derived from a Greek word meaning "circle of little animals." Although not all the symbols of the zodiac are animals, the Indo-European cultures developed a zodiac of twelve signs or constellations of stars through which the sun appears to pass during its yearly journey across the heavens on what is known as the ecliptic.

The twelve constellations are thirty degrees apart. Each symbolizes a particular phase of the sun's-hero's mythological story and impacts subconsciously this planet and its people. Starting with Aries, the Ram, the constellations alternate between being either masculine-positive or feminine-negative; there are six of each polarity in the zodiac. Each of the twelve constellations is also aligned to one of three qualities essential for spiritual evolution. These qualities are deemed to be either mutable, fixed, or cardinal.

Mutable: Gemini, Virgo, Sagittarius, Pisces
Fixed: Taurus, Leo, Scorpio, Aquarius
Cardinal: Aries, Cancer, Libra, Capricorn

As you can see, each quality is represented by four signs, which together create a cross. According to Alice Bailey, who channeled the sacred and alchemical teachings of the Tibetan Master Djwhal Khul,

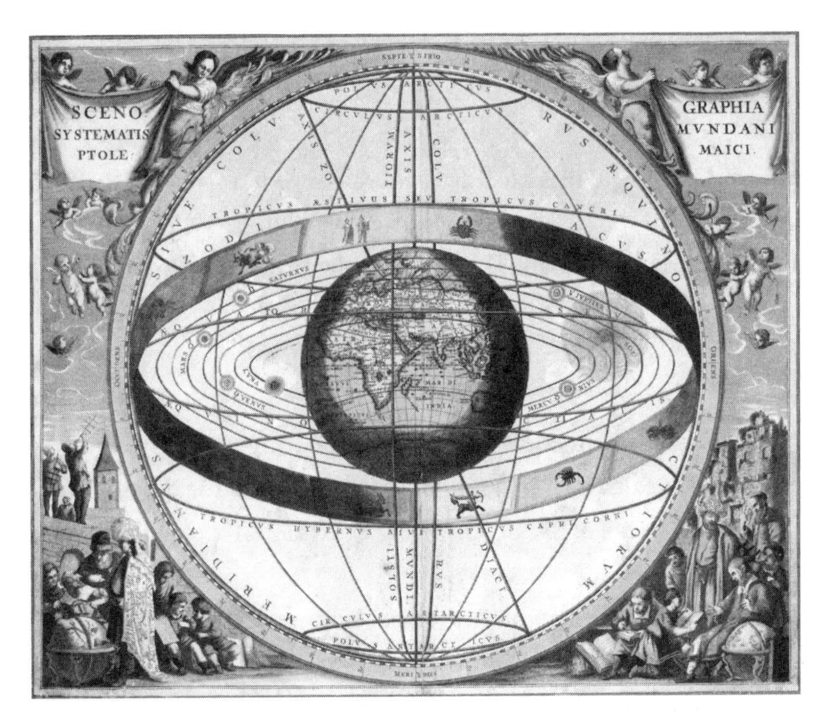

Map of the heavens (Andreas Cellarius, Atlas Coelestis Seu Harmonia Macrocosmica, *Amsterdam, 1660)*

our spiritual quest requires us to embody each of the crosses symbolically in order to reach spiritual enlightenment.[1]

The esoteric meaning and symbol of each cross is as follows:

The mutable cross: a place of action, reaction, experience, and the awakening of consciousness. The symbol is the swastika, which is an ancient sign reflecting the freedom that comes when we learn to master our thoughts and desires. Each zodiac sign of the mutable cross represents a different focus for the mind:

Gemini: the consciousness of duality
Virgo: the blending of spirit and form through self-reflection and contemplation
Sagittarius: focused awareness on the truth

Pisces: blended radiance through universal consciousness, which allows the individual to ascend to the fixed cross

The fixed cross: a place of transmutation where desire shifts into aspiration and selfishness into selflessness. The symbol is a cross whose four arms are of equal length. Each zodiac sign included in the fixed cross reflects the type of resources that will nurture our spiritual journey.

Taurus: the wealth that lives within, awaiting manifestation
Leo: the wealth and richness of the manifested form
Scorpio: the wealth of experience extracted through trials and tests
Aquarius: the wealth of group consciousness as the essence of light is finally extracted from the manifested form, allowing the individual to ascend to the cardinal cross

The cardinal cross: a place of transcendence where the individual is free from the restraints of the physical world and is in service to universal consciousness. The symbol for the cardinal cross is an X, with each zodiac sign in the cardinal cross symbolizing the creative abilities inherent within the individual.

Aries: carries latent creative consciousness urging the individual to take the risk to create
Cancer: the creative plan in manifestation moves the individual to the awareness of his or her ability to manifest his or her own reality
Libra: the creative plan realized in consciousness challenges the individual to use these gifts wisely
Capricorn: the transformation of conscious awareness through the perpetual fire into pure essence; the individual becomes his or her own initiator as the mystic

It is important to note that the statements above do not relate to our sun sign per se or necessarily to the placement of any of the planets within the natal chart. Instead, they relate to an esoteric understanding of spiritual achievement; each cross leads us toward our spiritual goal of eternal life. Yet it is true to say that humanity as a whole is, at present, poised on the fixed cross ready to ascend the cardinal cross through the portal of the age of Aquarius, which is just now beginning its twenty-one-hundred-year cycle. This epoch represents a time of self-responsibility, group consciousness, social conscience, and detachment from personal desires and aspirations.

THE PRECESSION OF THE EQUINOXES

Poised as we are between two great epochs of time, the age of Pisces and the age of Aquarius, this is probably a good moment to explore the planetary signification of such a transition. To ancient peoples, these changes were extremely important in that these peoples were deeply aware that the movement of the celestial bodies impacted our consciousness through the downloading of archetypal energy. It may surprise some readers to know that such astrological knowledge is present within the most widely read book on the planet: the Bible. Deep within its pages we find mention of an astronomical and astrological process that can transport us forward into the future or back into the past, revealing insights into our ancestors far beyond the reach of most historical texts.

In essence, these teachings relate to a process called the *precession of the equinoxes.* This describes a slow "wobble" of the earth on its axis. This wobble occurs over a period of twenty-six thousand years and occurs in the direction opposite the normal rotation of the earth. It is because of this wobble that the North Star, Polaris, is the star around which all other stars appear to orbit. This will not always be the case, however, for over time, due to the wobble, Vega will become the North Star (although this will not occur for many thousands of years).

Movement of stars around the North Star

Astrologically, approximately every twenty-one hundred years the alignment of the earth to the celestial equator changes, heralding a new age or epoch and, with it, a new frequency of consciousness for humanity. The fact that there are references to these different epochs in the Bible clearly indicates that its authors understood the nature of the precession of the equinoxes. At present, with the transition from the age of Pisces to the age of Aquarius, we are feeling the influence of both archetypal energies, with the former causing a large proportion of the population to give away their power to a higher authority, while the latter reminds us that we *are* the authority.

The Age of Pisces the Fish (0 CE–2100 CE)

Over the past two thousand years, under the influence of the fish, we have seen an expansion in religious and spiritual matters, mysticism, and creative endeavors. This naturally spawned the appearance of many gurus and leaders who quickly attracted followers. These men and women were only too happy to offer up their power and avoid self-responsibility. This could have been an epoch that saw the unification of humanity, yet it has probably been the bloodiest two millennia in the history of humanity. Factions formed around these gurus and leaders, and their battles still rage around the world today.

Because this age birthed Christianity, many Christian principles are based on the archetypal energies of Pisces. The fish is a common representation of the Christian faith. Jesus chose his disciples from fishermen, and his miracles often involved water, as revealed to us in the stories of his walking on water and turning water into wine.

Let us hope that as this age draws to a close, we will be able to see through our delusions and face up to our addictions and truly understand the unity that comes with the appreciation of diversity.

The Age of Aries the Ram (2000 BCE–0 CE)

This age began with the great shepherd Moses leading his people out of the land of Egypt. Yet the people had not finished with the old epoch, for once Moses' back was turned, they created a golden calf, asserting their connection to the previous age of Taurus the Bull.

Many religions conceived in this period still believe that the ram or sheep is sacred. Indeed, these animals often appear in many people's mythological stories. This age also saw the beginning of the patriarchy, moving away from the Taurean-based matriarchy and bringing with it an increase in war. (Aries is ruled by Mars, god of war.) At the end of the age (two thousand years ago), it was the shepherds who were the first to welcome the infant Jesus, handing over humankind's consciousness from the age of the ram to that of the fish.

The Age of Taurus the Bull (4000 BCE–2000 BCE)

This age inspired mainly matriarchal and agricultural societies as the people released their hold on the nomadic lifestyle found in the age of Gemini; they were willing to take guardianship of the land. It was out of this age that both Judaism and Islam were conceived through their father Abraham, although in truth it was Sarah's insistence to have a child that led to the birth of these two great religions and their subsequent disharmony.

During this time, Hathor, the heavenly cow goddess, reigned supreme in Egypt, promising abundance to her people. All religions conceived during this time still consider the bull to be sacred.

The Age of Gemini the Twins (6000 BCE–4000 BCE)

Little is known about this age although it can be assumed, through our knowledge of astrological symbolism, that during these years the consciousness of the people would have been focused upon the intellect and the development of increasingly advanced communication skills. Nomadic existence would have also been common, and the cross-pollination of ideas, philosophies, and language would have existed across the globe.

The Age of Cancer the Crab (8000 BCE–6000 BCE)

As with the previous age, there are few signs that remain to inform us of the consciousness of the people during these years. Yet, it is clear that the principles of the feminine would have taken precedence; home, nurturing, and family were major priorities.

The Age of Leo the Lion (10,000 BCE–8000 BCE)

We do know something of the age of Leo, for an artifact from that epoch still stands: the Sphinx. On the plateau of Giza, clear for all to see, we find a figure that is half human and half lion and is said to symbolize Sehkmet, the lion goddess of the time. Known to be much older than the pyramids and made from natural rock, the Sphinx is scarred by the presence of salt water. This tells us that it was already carved when the area was last surrounded by seawater some nine thousand to ten thousand years ago, during what is known as the Great Flood, which is associated with the building of Noah's Ark. During this age, personal power and authority were paramount and humanity experienced itself as king of the jungle.

Now, as we are about to enter the age of Aquarius, the zodiac sign that sits directly opposite Leo, we may wonder what our ancient ancestors would say of our progress. One clue of the times ahead was left by Jesus. When his disciples asked how they should find him, he replied:

The Sphinx

"Follow the man who carries the water" (Luke 22:10). This water carrier is the symbol for the age of Aquarius and reveals that the unified field of Christ consciousness will be found not in a single man but in everything on this planet that contains life, during the new world of the fifth sun.

5

THE RADIANT BLUEPRINT

Let us now begin our journey, recognizing that the twelve steps we are about to embark upon have been taken by mystics for thousands of years in their search for eternal life. Understandably, our starting point is the ocean of possibilities where everything awaits manifestation. Just as we are presently suspended between the fourth and fifth worlds, so we experience a momentary pause between each breath when all our options are potentially open. The next in-breath, or inspiration, brings focused attention upon the ocean of possibilities, and a whole new cycle begins.

This state of dynamic potentiality is reflected in the archetype of the Virgin whose name, contrary to popular belief, connotes not an innocent, sexually unaware girl, but instead means "to be complete unto oneself without the need of another to make one whole." Often depicted wearing white, the Virgin radiates a state of completeness awaiting manifestation and experience. An analogy is that when we buy a jigsaw puzzle, the picture on the front of the box reveals wholeness, and yet it is only through working patiently with each piece that the picture is recreated from the building blocks within.

The Virgin can be described as:

- Our higher self, which does not incarnate within the physical form

- Our spiritual blueprint awaiting manifestation
- Our akashic record, which chronicles the consciousness of our existence
- Our unrealized or unmanifested self
- Imagination containing all the seeds of possibility
- Collective consciousness awaiting attention
- Energetic perfection
- Our holographic self in which everything is essentially present

The astrological sign associated with this dynamic state of potentiality is Pisces, symbolized by two fish swimming in opposite directions and joined together at their center. This is a reminder that here in

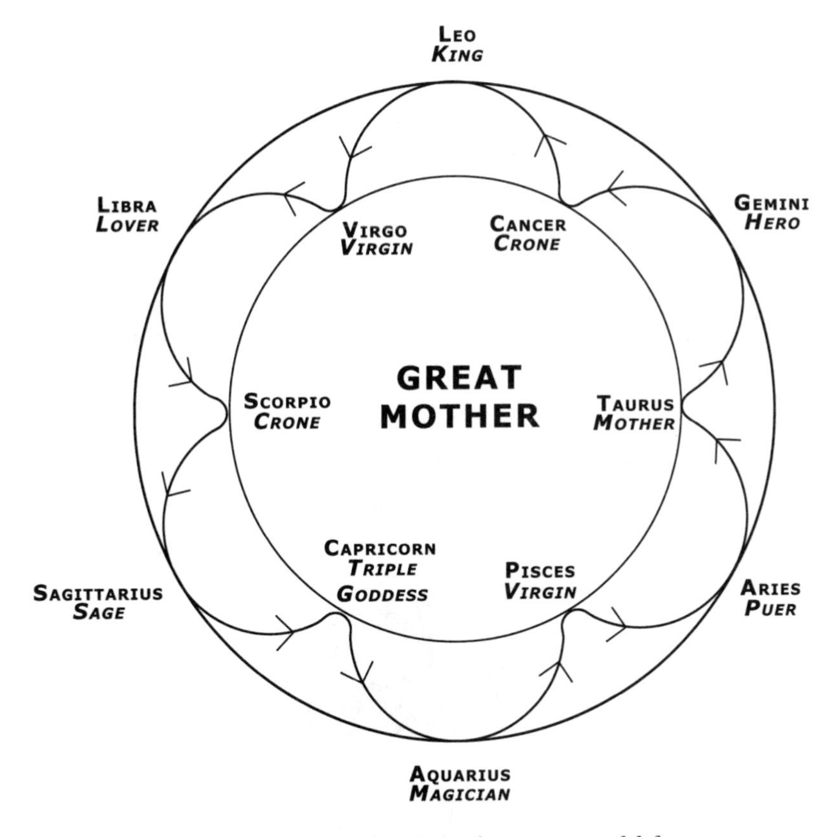

The creative cycle leading to eternal life

Pisces we are poised between the worlds of possibility and probability wherein the focus of our attention is the catalyst that drives the scenario in either direction.

This is the world of impermanence and opportunity as long as we remember that we can enter the unified field of potentiality only through understanding the principles of duality.

♓ PISCES

Quality: mutable; merging with universal consciousness

Alchemy: Coagulation; one with all consciousness

Polarity: feminine; Virgin, higher self, the unmanifested self, being complete within oneself

New Moon Phase: to germinate and emerge

As with the phases of the moon and the description of the moon tree, we see these two fish held in dynamic tension by their joint energies, which both attract and nurture at the same time that they oppose and destroy. This is the same principle represented by the yin and yang symbol in which the "eye" of each fish is the seed of the opposing force that will be nurtured by the demise of the body.

Between the two fish in the sign of Pisces is the link symbolizing the fluidic central force that is the source of their existence and is fed by their interaction. This symbiotic trinity is a theme that repeats itself throughout our journey, reminding us that nothing and nobody upon this planet can exist separate from the rest; all of us are interdependent.

Yin and yang symbol

Hence, the trinity can be symbolized as:

- **The Triple Goddess:** the duality represented by the Virgin and the Crone with the central beam of the Mother
- **The Christian Trinity:** the duality represented by the Father and Son with the central beam of the Holy Spirit or Mother
- **The moon:** the duality represented by dark and light with the central beam of the sphere or vessel from which both appear and disappear

To summarize: Immortality is a dynamic movement of energy between different dimensions of existence that feed and are fed by an eternal fountain of life.

THE ETERNAL HISTORY KEEPERS

There is one animal that epitomizes the essential qualities described above: the whale. This mighty mammal circumnavigates the vast oceans of this planet and within its sonar energy field it carries records of our collective consciousness or, in other words, the akashic record of the world. We have only to read the story of Jonah being swallowed by a whale (Matthew 12:40) to appreciate that this powerful mammal is capable of amazing feats of alchemical transformation that take place every second as it sends its sound waves through the medium of water. Symbolically, Jonah's three-day descent into the whale's belly represents his immersion in his own akashic record, or spiritual blueprint, causing him to emerge a completely changed man.

The parallel for humanity is that we too must be swallowed metaphorically by our higher self, reconnecting to our spiritual roots and evolving to the next level of consciousness in readiness for the new world to emerge. We are not alone in this process, for the whale constantly sends out sonar webs of creativity into the oceans of possibilities to help us remember our inherent connection to the Great Mother.

Humpback whale

Each human consists of 70 percent water, thus we cannot fail to hear these messages, even though they may reach us subliminally or be in a language that only the heart understands. It is interesting to note that over the past few years, the songs of the whales have been changing, aligning us to the new frequencies of consciousness that resonate with the transformation of the earth and her inhabitants.

Unfortunately, within this same time period, there has been an increase in the practice, by the navies of the United States and England, of sending extremely low frequency (ELF) sound waves into the seas, ostensibly to enhance communication between submarines. The result has been an increase in the number of beached whales and dolphins due to damage to their sonar systems. Could it be that those who have the authority to order such maneuvers are attempting to disrupt the communication between these wonderful mammals and ourselves? Could it be that certain powers prefer that we do not remember that we are first and foremost powerful eternal beings?

THE VIRGIN: RADIANT PURITY

As we deepen our study of the Virgin within mythology and her relationship to wholeness and purity, it is clear that one of her representatives in recent times within the Christian world is the Virgin Mary.

Unfortunately, Mary's image has been exalted to such an extent that many of her followers have become lost within its shadow, but it is important to understand that her quintessence of purity—contrary to popular belief—is related not merely to human goodness, but also to an authenticity from which the veils are removed. Nothing is concealed, and she radiates, to the fullest of her being, pure white light.

With this distinction in place, it is much easier to acknowledge that we already carry the blueprint of spiritual perfection or wholeness within us which, like the jigsaw cover, is merely awaiting manifestation. Now it is easier to understand why the archetype of the Virgin is often portrayed as being unmarried, for it is more accurate to say that she has no need of another to make her whole. Picture a world in which our relationships are not bound by the fear of rejection or the neediness of others in order to know completion. Imagine the joy of just being in the presence of another soul who knows and trusts his or her implicit wholeness.

Such an expression of this pure maiden archetype within our society is relatively uncommon, although it can certainly be witnessed in the crystal children who embody the principles of the etheric fifth world. These young people are strongly connected to the source of their inspiration and hence to their spiritual blueprint and thus appear not to require parenting in the traditional sense. They appear self-assured and detached and yet radiate a compassionate presence. We can only imagine how the appearance of relationships will change as these children enter adulthood. Being "complete unto themselves," their interactions will be free of the stickiness that usually accompanies unspoken expectations and low self-worth.

On an esoteric level, the Virgin's lack of need for another to make her whole equates to the fact that, dwelling as she does within the unified field where there is no separation, she has no concept of *another* but only the appreciation of *us*.

Brigid: Keeper of the Fires of Creativity and Fertility

While honoring the essential and energetic completeness of the Virgin, we also must recognize that this energy of potentiality seeks manifestation in the world, which is symbolized by the Virgin giving birth to the boy-child. This concept is epitomized by the young girl whose seeds of creation—her ova—are in place fully at the time of her birth, merely waiting the moment of fertilization.

Thus we meet Brigid, one of the oldest of the Celtic Virgin goddesses. Her impact on the cycles of creativity and fertility is still remembered at Imbolc, the festival of the lactating ewes that heralds the return of the life-giving forces of spring on February 1 and 2 of the Celtic Calendar. She has been known by many names, including Bride, Brigantia, Brigit, Bridey, and Bridget (her Christianized name), which all mean the same thing: "fiery arrow of power."

There are suggestions of her cultural influence before the Celtic period, possibly dating as far back as the building, in England, of the megalithic sites of Avebury and Stonehenge, whose massive stones are known as bridestones. Other tales tell how Cailleach, the Scottish name for the divine Hag or Crone, imprisons a maiden named Bride in the high reaches of Ben Nevis. When Cailleach's own son falls in love with Bride and they decide to elope at the winter's end, Cailleach sends fierce storms to prevent their union. Eventually, love prevails when the Crone is turned to stone, leaving the couple free to enjoy their life together.

This story shows the beautiful balance between the Virgin and the Crone, highlighting Cailleach's rule over the winter months and Bride's or Brigid's dominion over summer.

In Celtic mythology, Brigid is the daughter of the Irish god Dagda, famed for his great power; his cauldron of eternal abundance (also known as the Holy Grail), and his magical oaken harp, which brings the seasons into order. Many of these skills are passed down to his offspring. Brigid is here seen as the goddess of the fire, her hair shining like the golden rays of the sun.

As mentioned in reference to the moon cycles, many ancient fires were dedicated to Brigid, the most famous of which are found in Kildare, Ireland. Yet her fires of continual fertility and creativity were also symbolized in other ways:

- The fire of the hearth, representing the center of the home or community from which all subsequent fires are lit, ensuring continuity of creativity, prosperity, and abundance. These perpetual fires, representing our heart, reassure us of our own eternal abundance as long as we nurture it joyously with love.
- The fire of inspiration, seen especially in poetry and song in which the words carry the frequency of pure insight and intuition
- The fire of healing found at sacred wells. There are many such sites around Britain that are dedicated to St. Bride or St. Bridget. These waters, rich in minerals, create the perfect balance between the fire and the water used in the process of purification and alchemical transformation or healing. Over time, many of these wells have become covered over or hidden; only now are they reappearing. Some of the most famous wells are found at Cullin near Mullingar, at Bride's Mound in Glastonbury, and at St. Bride's Church, London.
- The sexual fire linked to the serpentine kundalini energy and seen as the inner fire that brings us to the state of self-realization.
- The fire of transformation that mirrors the skills of the blacksmith who bends and twists metal into a creation of beauty. Brigid as a master alchemist uses her sexual (kundalini) energy to turn base consciousness into the golden essence of illumination.

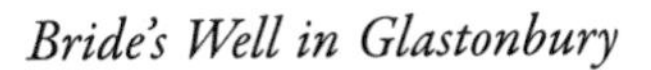

Bride's Well in Glastonbury

Brigid's cross

The synthesis of these fiery qualities is, to this day, symbolically woven together in a cross dedicated to Brigid. It is created from twisted stems of barley or corn and shows up at the festival of Imbolc. Its specific design naturally expresses movement that reflects the birth of a new cycle of life as consciousness spirals to the next level.

Pallas Athena: The Keeper of our Wisdom

The next face of the Virgin is as wisdom keeper or our higher self, guiding us on our journey back to wholeness. In mythology, these qualities are expressed by the Virgin goddess Pallas Athena, whose roots stretch far earlier than the Greek civilization, emerging originally from the element of water, similar to many of the great maiden goddesses. It is even thought that she lived among us as an enlightened being or ascended master during the Lemurian civilization some one hundred thousand years ago.

Athena represents the principles of truth and wisdom which, contrary to the opinions of pragmatists and fundamentalists, are not set in stone but instead emerge out of our own unique creative experiences within the holographic universe. Athena speaks to us through our intuition. This still, small voice of inner reason or knowing maintains the connection to our spiritual blueprint, urging us to fully manifest this blueprint in the world. Then, when the time is right, she is there beside us, compelling us to allow the old to die so that the essence of the expe-

Pallas Athena

rience can be extracted as light, transporting us to new levels of awareness and the creation of a new blueprint.

With the advent of the Greek civilization and its strong patriarchal and intellectual bias, it was determined that if Pallas Athena was to continue to guide her people, her persona must be altered to become equally acceptable to both men and women. Thus she was no longer born from the waters of the ocean but from the head of her father, Zeus, fully armored, protecting her femininity, and prepared to channel her wisdom through the mind.

Over time and despite the masculine tendency to solidify wisdom into dogma, the goddess kept the timeless and essential quality of truth alive through the arts, intuition, and inspiration. Yet her hold was tenuous and her inner knowing was often deemed to be "feminine fancy," which was perceived to have arisen from the chaotic realms of an unstructured subconscious that must be controlled at all costs.

Fortunately, Jung and other great psychologists arrived on the scene just in time to bear witness to the archetypal energy of Pallas Athena. Through their research, they showed the world that spiritual evolution is limited severely when viewed only through the lens of the logical mind. The door to the return of Pallas Athena was opened, and today we see her standing comfortably, complete in her own nakedness and without the need for mental defenses. She calls on all of us to follow her, to think outside the box, and to appreciate that what we call the unconscious or imagination is, in fact, the ocean of possibilities from which all dreams arise and all wisdom will return.

Probably one of the toughest issues in these times is that despite the fact many are choosing to awaken to their potential, there is no one route to follow, no particular path to take, and no one master or messiah to tell us what to do. The old age of Pisces is dying. Then, we could always rely on a guru or leader to show us the way. Now, as we face the age of Aquarius, we are being asked to develop self-consciousness; each of us must accept responsibility to fully embody our incarnated soul and hence together move forward as a collection of highly individuated and connected beings.

Such a shift from giving away our power to an authority figure to being accountable for its effects is not easy. We have become comfortable looking outside ourselves for solutions, encouraged by those who subversively perpetuate the pattern in order to ensure that their own power and our dependency continue. Yet times are changing, and despite the inconvenience of having to take responsibility for our creations, there is a relief in not having to:

- Spin plates to keep everybody happy
- Wear masks to avoid being seen fully or upsetting others
- Stay small or inside a box where the sides dig painfully into the skin
- Be the proverbial round peg in a square hole
- Carry ancestral baggage, fearful that if we lay it down, there will be no purpose for our life

- Do something we do well but which no longer makes our heart sing
- Convince ourselves that tomorrow will suffice when it comes to following our bliss

Now is the time. This is the place. We have arrived . . . and the Virgin is ecstatic! Holding the blueprint of our incarnation, Athena is our constant companion, urging us on when we become paralyzed with fear, sitting with us when we are weary and in need of rest, and celebrating with us when we resonate with the light of our true nature.

Intuition is more than just an intellectual knowing. It is a vibration that resonates with the pulse of our soul and excites the dream into reality. Despite this natural mechanism of creation, there are many potential diversions that can cause an idea to fail to reach manifestation. These diversions are fear-based, including fear of failure, fear of the unknown, and fear of losing control. Yet the Virgin is patient and will wait until we realize that:

- There is no judgment of our actions except within our own minds.
- The tried and tested usually leads to stagnation, while the unknown leads to an ocean of possibilities.
- True control comes from following the unpredictable call of the soul.

Pallas Athens entreats us to follow what is most nurturing to our inner being, rekindling the eternal fire and moving away from situations based on fear that separate from us from the blueprint of our heart's desire. She offers us this one simple question to be applied to all our actions: Am I acting from love or from fear? Love connects; fear separates.

As we will see, Athena's presence never leaves us. Her truth is always the same. She encourages us to follow that which excites, nurtures, and connects us to our heart's desires.

6

FROM BOY-CHILD TO KING

The next five phases of the spiritual journey describe the growth of a dream or idea as it pushes its way from its resting place, strengthens its core, and ultimately stands tall and proud, king of all it surveys. Together, the stages represent the development of the first length of the magician's wand, known esoterically as the *ida,* which runs along our spine from base chakra to crown.[1] During its development, energy is transformed from the higher frequencies of spirit into the denser frequencies of matter as the child becomes king.

At this critical time in our history, this half of the cycle is, in some ways, less important than the final stages that ask for the king to lay aside his crown, face his demons, and eventually die back into the ocean of the Great Mother, relinquishing all sense of separation. Yet our ability to fulfill our spiritual destiny is predetermined by the strength of

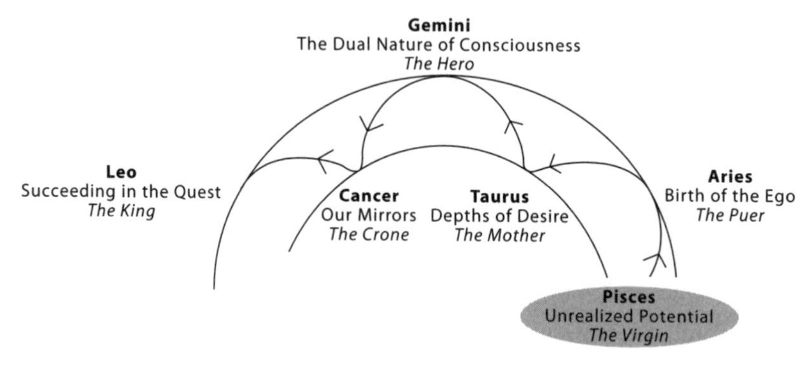

Five stages of inspiration

our successes, for it is this energy that will sustain us during our passage through the underworld.

Most of us have at one time or another tasted the sweet nectar of success and felt the surge of joy when we see our creative ideas come into realization. Yet we have dreams that still await manifestation but have been abandoned due to a fear of the unknown, a lack of encouragement, or a failure to appreciate the true meaning of success. We may have chosen to store these unfinished projects in convenient closets, believing that someday we will complete the cycle. That day has arrived, reminding us that to die with regret is one of the greatest expressions of sadness for the soul.

At this present time, we can see few individuals who are creating anything new; most are merely completing tasks that have languished untouched for years and often lifetimes. Thus even a new partnership may exude an aroma of familiarity as we reconnect with a soul friend after hundreds of years and choose this time to see the relationship through to the end.

It is time to open those boxes and closets and decide what can be completed to the point of soul fulfillment and what must be allowed to die because it no longer nurtures the soul. As we take on this challenge, it is important to include our relationships, goals, jobs, and "to do" lists, determining what brings light and joy and what feels like a burden.

To help in the "spring cleaning," each stage of inspiration will be discussed in turn, with the purpose of jogging the memory into action and elucidating why the situation or scenario was created in the first place as we remember that, fundamentally, we are the creator of our perception of reality.

THE CHILD OR EGO IS BORN

The Virgin is ready and her eggs of potentiality are awaiting the sperm, but what is the impetus that causes us to fertilize our ideas, start a fresh project, decide to take a risk, or seek new pastures to explore? The

Virgin may offer intuitive insights from our blueprint and the energy of creation, and yet she is not the one who decides to focus attention, initiating a whole new cycle of creativity. Many have been taught that God makes such decisions. They give away their authority by saying, "It is not for me to know or understand but just to do. Thy will, not mine."

Yet there is an instigator within us that knows how to initiate creativity by focusing on the ocean of possibilities and causing the Virgin's energy to begin its transformation into matter. This is the alchemist's One Mind, symbolized by the determination and willpower of the Ram associated with the astrological sign of Aries. This call to adventure takes place long before we emerge from our mother's womb and is recorded universally at the moment our soul agrees to incarnate on Earth at this particular time. Few individuals can recall this moment preconceptually (and the majority of the population is still trying to understand how they were hoodwinked into accepting such an offer!).

♈ ARIES

Quality: cardinal; latent creative consciousness and the power to create

Alchemy: calcination; stepping beyond the confines of the collective

Polarity: masculine; puer, adventure, curiosity, and innocence

New Moon Phase: to germinate and emerge

Despite the fact that we may wonder about our acceptance, it is important to remember that we did agree to this adventure, acknowledging that it is a unique opportunity for soul growth and transformation. Thus we have created the perfect scenario, calling in family members, friends, partners, societies, nationalities, genders, and even enemies to assist in bringing alive our spiritual blueprint. It could be said that some of us made strange choices, but that is for us to know and not for others to judge!

It is good to appreciate that the puer stage of the cycle is not confined to our physical birth, for we actually pass through this phase many times in one life. On each occasion we emerge from the Great Mother's ocean like a newborn child eager to start again and perhaps do things differently based on the wisdom of experience. Under the protection of the Virgin, we are motivated to leave the safety and comfort of the known and seek our own unique individuality. This is also known as the development of the ego.

It is the Virgin who drops the seed of an idea into the chalice of our mind, knowing that a certain amount of time will pass before the idea sees the light of day. It is her voice of determination that calls on us out of the blue to rise up and leave a humdrum life and enter one of colorful excitement. It is the Virgin who whispers in our ear at the moment of our waking, causing us to exclaim, for no explicit reason: "Today is the day!"

Despite the clarity of the call, the true purpose or outcome may often be concealed as if to test our faith. Indeed, the very fact that I live in the United States resulted from an innocent question that someone asked of me soon after the death of my mother. I heard myself say with certainty, "I am going to live in America," and because I seemed so certain, nobody questioned my decision. Responding to an intuitive insight might sound extremely reckless to those who would never leave home without adequate financial resources and a specific plan of action, and yet it is this mix of excitement and fear that strengthens the puer until he becomes a man.

Our inner guidance knows the type of "candy" or "carrot" required to induce us onto the path, certain that once our commitment is guaranteed, any change in the agenda is unlikely to throw us off course and we will gladly move forward. Thus we may remember a time when the stranger we sat next to at a seminar was more influential than the speaker we paid to hear, or we might recall following the love of our heart to another state or country only to fall out of love with the person and in love with the location. In other words, it is not always important that the puer knows why or where he is traveling. What is important is that he agrees to the adventure.

Planetary Momentum

Astrologically, there are specific times in our lives when we experience a heightened inspiration to follow our destiny. This is particularly true as it pertains to the planet Saturn, which orbits the skies approximately twenty-eight to twenty-nine years before returning to where it was on the day an individual was born. Hence, the years when we are twenty-eight or twenty-nine, fifty-six or fifty-seven, and eighty-four or eighty-five are all weighted with the impetus to ask, "Am I living my soul's desire? Am I following the blueprint of my life?"

For some people, twenty-eight years is too long. Instead, they prefer to review life every seven years, giving rise to the notion of the seven-year itch. Another notable age is thirty-three, when the soul begins to fulfill its promise of service to the Great Mother. With this in mind, it is not uncommon to hear that many ascended masters, teachers, and prophets began their true work at this age. Some individuals may choose to leave the planet at thirty-three, their inner work having been completed.

Choices: Acceptance or Refusal?

A common point of discussion is whether our lives are destined or subject to free will. Perhaps both are true. It is fate that sends out the messages at certain times in our life and free will that allows us to

- fail to hear or conveniently forget the message,
- sabotage any efforts to strike out on our own unique path,
- set unreasonable conditions before we take the first step, concealing an underlying fear of change, and
- see an opportunity for soul growth and service to the Great Mother.

From an intuitive perspective, our guides within the world of spirit cannot override our free will or judge our choices. Yet this does not prevent them from doing everything in their power to attract our attention and then shower us with encouragement as we heed the call to adventure.

So as we seek to fulfill our soul's destiny, it is important to ask:

- What dreams from my youth have not been realized?
- What excites me, filling my heart with joy?
- What would I regret never having the chance to achieve?
- What would I do if my life wasn't driven by *should, must,* or *ought to?*
- What inspiration has been filtered down to me from the Virgin and still awaits manifestation?
- If I loved myself unconditionally what would I do?

As a woman once asked me in a lecture: "Jesus came to me five years ago in a meditation, gave me a brush, and said I should paint. What should I do?" If the masters take the time to visit, please do not wait another twenty-six thousand years to manifest your destiny!

DIVING INTO THE DEPTHS OF DESIRE

In order to fulfill his ideas and dreams, the puer must focus his attention on the abundance of the Great Mother and draw toward him all those things that will nurture him on his journey. This phase, under the watchful guidance of the matriarchal energy of Taurus the Bull, reveals to us a wealth of skills, talents, and gifts, which ultimately underpin our success as king.

♉ TAURUS

Quality: fixed; inner wealth awaiting manifestation, desires

Alchemy: dissolution; discovering the gifts and talents that lie within

Polarity: feminine; Mother, nurturing possessions, skills, and talents

Crescent Moon Phase: to move and focus

The issues that emerge at this stage of the journey include a failure to surround ourselves with the skills and nurturing required to support our creative impulses. This is due commonly to a poor sense of individuality, causing us always to feel a sense of poverty and scarcity in our lives. Another issue is stubbornness; the desire to possess becomes more important than the desire to create, causing us to hoard or bury our talents.

What is vital to remember is that there is no purpose in possessions unless they nurture our dreams toward reality. Once this reality has been achieved and we become king, they become surplus and are taken from us as we descend into the underworld to embrace a much more powerful wealth that can never be possessed.

Meeting Our Mentor

During this phase it is not uncommon to encounter a character who offers wisdom and guidance for the journey ahead. Part Virgin and part Crone, such an archetype is perceived as a mentor and, in traditional stories, is dressed as an old man or woman, symbolic of the archetype of the sage, mystic, or hermit. Such archetypes usually appear to be "in the world but not of it" and hence may accompany a traveler for a while before disappearing or dying. Thus we meet Merlin in the Arthurian legends, Gandalf in *Lord of the Rings,* Obi-Wan Kenobi and Master Yoda in *Star Wars,* and the fairy godmother in the *Cinderella* tale.

Mythological mentors often possess a magical and mysterious quality, leaving us with little doubt that their supernatural qualities are a result of their own internal battles and the eventual mastery of their energies. Such strength reminds us that those who step up as mentors must be dedicated to the task and selfless, unattached to a specific outcome and not seeking self-gratification.

I have been privileged to meet a number of fine spiritual mentors. Some walked with me for mere minutes and some supported my journey for years. Not all were comfortable to be around, and yet there was no doubt that they loved me, and I certainly wouldn't be where I am now without them. Three of the most significant mentors were men, each dying suddenly, seventeen years apart. The first was my loving and adventurous father, who protected and supported my physical and human nature until I was ready and able to take care of myself. Seventeen years later, I met a great love who opened my heart to my emotional and transpersonal self; he died tragically soon after our meeting. He came into my life for a purpose and left when he knew his work was complete. Finally, I was privileged to meet a wonderful kahuna, or wise man from Hawaii, who joyously supported my spiritual being for three years before being killed in a car accident. His contribution to my soul completed the mentoring of my body, heart, and spirit. I am blessed to have been acknowledged, encouraged, and loved by these three amazing men, all of whom still watch over me from beyond the veil.

The final tools that the puer extracts from the Great Mother are the twin forces of spiritual and physical strength, transforming the boy into a hero as he learns to master these essential forces of creation.

THE DUAL NATURE OF CONSCIOUSNESS

Probably one of the greatest challenges we encounter during our spiritual evolution is mastery of our own power. The first time we meet this issue is in the sign of Gemini, as the hero takes on the strength of the twin powers. These powers are represented by the physical-masculine and the spiritual-feminine faces of the Divine, both of which must be acknowledged equally and mastered. In terms of consciousness, the masculine is seen to pertain to intellect while the feminine represents intuition.

♊ GEMINI

Quality: mutable; the consciousness of duality

Alchemy: separation; owning one's own power to make decisions

Polarity: masculine; hero, strength and resourcefulness

First Quarter Phase: to build and decide

It is vital that the hero appreciates that each power is only as strong as the relationship it has with the other; the natural tension among them leading to reciprocally rewarding growth. This is the same dynamic that exists between the two fish of the astrological sign of Pisces: Their natural friction supports and is supported by the link connecting them.

In other words, as long as the hero maintains a healthy respect for both his spiritual-intuitive and physical-intellectual natures, then both will serve him well, nurturing his eternal being. Should either take priority, however, or should the flow between them become blocked or broken, then our hero's journey will be limited both in terms of material success and spiritual fulfillment. This being said, it is valuable to see a reflection of such events woven into both ancient and modern history.

The Pillars of Jachin and Boaz

The symbol for the sign of Gemini is believed commonly to represent the twin pillars of Jachin and Boaz, which are fundamental to the principles of Freemasonry, symbolizing establishment and strength, respectively.[2] The history behind these structures dates back to the land of Egypt, when, prior to the building of pyramids and ziggurats, pillars were considered to be a potent means of linking the worlds of men and gods.

Such pillars or obelisks were seen to symbolize the fully erect serpent whose magical qualities, like the wand, act as a lightning rod, bringing heaven onto earth and earth into heaven. As we are beginning to appreciate, our body is also designed to be a magician's wand, with three strands of energy running along the spine and contributing to the fully erect serpent: two forces of duality and the fluid source of eternal life that flows between them. These are named the *ida,* which flows from base chakra to crown, the *pingala,* which flows from crown chakra to the base, and the *sushumna,* which gives life to and absorbs the outer energies, representing the continuum or the nothingness of the Great Mother.[3]

Lightning rods

The symbol used most commonly to represent this trinity is the caduceus, wherein two serpents are seen to wind around a single rod, their fully developed wings rising close to the top. Often confused with the single serpent of Asclepius, the god of healing, the caduceus is the staff of Hermes, messenger of the gods and ruler of the sign of Gemini. Once we appreciate that the Greek Hermes is synonymous with the Egyptian god Thoth, the author of the Emerald Tablet, it becomes clear that this ancient alchemical symbol depicts the activation of our energy channels until we have the "wings" to fly beyond this three-dimensional world of reality. The position of these wings at the third eye (*ajna*), located between the eyebrows in the two-lobed energy center, and their relationship to the swan will be discussed in later chapters.

The caduceus

At the top of the rod of the caduceus is a small sphere that represents the pineal gland, connected esoterically to the crown chakra. Mystics have always understood this tiny gland to be the seat of the soul. Not only does it produce hormones such as melatonin and DMT, which allow our consciousness to connect to other realms of awareness, it is also capable of producing light when stimulated by the right frequencies. This knowledge corroborates the fact that when the three streams of energy flow up and down the spine, the pineal gland is awakened and stimulates the activation of the Ka.

With this esoteric understanding, the significance of these obelisks takes on a whole new meaning. Thus, when the peoples of Lower and Upper Egypt agreed to unify, each presented their own sacred pillar or divine connection to the union, agreeing that they should be linked by a central beam and thus establish stability across the land. The right-hand pillar, Jachin, came from Lower Egypt and represented spiritual authority (establishment), while Upper Egypt contributed the left-hand pillar, Boaz, symbolizing physical strength and intellect. Together, they created a doorway between the dimensions, facing east to meet the rising sun, a tradition continued in Masonic temples even today.

The central beam originally represented the sun god Nut but was later said to represent Ma'at, symbolizing righteousness, fairness, and justice. Such qualities arose from a place of detached compassion, for

Ma'at was seen to exist beyond the small minds of men and therefore was unbiased, nonjudgmental, and connected to that which was seen to benefit the greatest number of people. It was clearly understood that the political stability and prosperity of a country was dependent upon this trinity comprised of the two pillars and the ever-evolving energy of the central beam. Ma'at was seen to be both the product and the source of the dynamic harmony that existed between the dual powers of spirit and matter.

As the years passed, Ma'at was perceived as a goddess, with an ostrich feather in her headdress to symbolize her commitment to the earth. She carried the ankh, representing the key to eternal life. The Egyptians worshipped Ma'at, for without the order and truth she embodied, they knew their world would flounder in primal chaos or solidify and die. She was seen to be the female counterpart to Thoth, who brought science, medicine, writing, and magic to the people.[4]

Over time, the feminine connection uniting the twin pillars was lost and the crossbeam came to represent Yahweh, a storm god who, identified by the symbol T or Tau,[5] united the kingdoms of Judah and Israel under the rule of King David and then King Solomon. Indeed the famed and wealthy Temple of Solomon was entered through the same twin pillars and became the inspiration for the columns (inaccurately

Ma'at

known as the apprentice and master pillars) in Rosslyn Chapel, built by the predecessors of the Freemasons, the Knights Templar.[6]

With this change in the guardianship of the crossbeam, the unbiased authority that Ma'at embodied was lost and thus began the development of rules and threats to keep the people in order. At the same time, the pillars took on more practical definitions. Strength (Boaz) was seen as a quality of the king who looked after defenses, law, and government, while establishment (Jachin) was the premise of the priest who managed religious righteousness. Together, they supported *shalom,* which means "peace through the right balance of governance between church and state."

In Freemasonry today, the meaning of the trinity is much the same; the ideals of strength and establishment are fundamental to a prosperous and stable society unified through the perceived word of God. Indeed, the Constitution of the United States was created around this premise by our forefathers, many of whom were Freemasons. Yet unless the principle of the central beam is acknowledged and honored, the passage between the worlds will become obstructed and even closed, causing us all to forget that, first and foremost, we are eternal beings.

It is imperative to understand that the archetype of Ma'at does not call for an obedience to laws and rules that generate fear, guilt, and shame, nor does she ever subscribe to the idea of favoritism. Instead, she embraces all peoples equally in the name of justice and fairness. At the same time, both vertical pillars must be treated with equal respect, acknowledging the fact that our spiritual and physical powers are interdependent; one is unable to survive without the other.

Some authorities say that on the original Egyptian pillars there were intricate carvings representing secret teachings about immortality, while others believe the pillars were metaphors for the kabbalistic Tree of Life. Whatever their secrets, there can be no doubt that the occurrences of September 11, 2001, when the Twin Towers in New York were destroyed, were intricately tied to these ancient teachings and reconnected us metaphorically to the true meaning of the crossbeam and its link to eternal life.

Grasping the Head of the Serpent

There is one final issue that must be addressed at this level. To master these energies, the hero must, metaphorically, be willing to grasp the heads of these two serpents so that instead of allowing them to thrash about wildly, he can focus their power and bring it into alignment with his soul's intention. Unfortunately, many individuals do not understand this spiritual principle, enjoying the relative freedom and strength that comes with an awakening psyche, but not appreciating the need for healthy boundaries and self-control.

Thus we see the emergence of power games and manipulation, the moving, taking, and even "vampiring" of energy without conscious accountability for our actions. Ultimately, the hero who fails to control his energy fails to build a serpentine pathway strong enough to allow him to pass to the next level. This forces him to always be dependent on the energy of others to maintain his strength.

MEETING OUR MIRRORS

Inanna is a young Sumerian maiden (soon to become queen) who lives on the bank of the Euphrates River.[7] One day she sees a tree floating in the water; it has been uprooted by the wild southern winds. She takes the huluppu tree from the river and plants it in her sacred garden, longing for the moment when she can use its wood to create, for herself, a shining throne and a wonderful bed to lie upon.

Ten years pass and the tree is still not ready for use. Then one day, Inanna is horrified to see that her precious tree has become the home of a serpent that cannot be tamed. He lives in the roots, an anzu bird nests in the branches, and the dark maid Lilith lives in the trunk. Inanna weeps!

Eventually, she calls upon her valiant, mortal brother Gilgamesh, who cuts down the tree with his bronze ax, causing the serpent to slither away, the bird to fly to the mountains, and Lilith to flee to wild and uninhabited places. With the trunk of the wood and at her request, he fashions a throne and bed for the Virgin, and she is well pleased.

Now with the dual forces of physical strength and spiritual righteousness in his hands, the hero dives once again into the Great Mother, attracting all manner of trials and tests similar to those faced by Odysseus in the *Iliad* and, more recently, by the courageous and wily Indiana Jones.

While the hero would like to believe that his mission is to rid the world of evil, battle to save the lives of the innocent, and win the hand of the beautiful woman, in truth many of his adversaries are mirror images of his own, hidden qualities, drawn into his awareness in order to strengthen his inner core. Just as the pincers of Cancer the crab meet in the middle of its body, so the hero attracts opposite poles of existence into his life so that he can learn to live comfortably within either pole while always maintaining his center.

♋ CANCER

Quality: cardinal; the creative plan in manifestation

Alchemy: conjunction; recognizing and accepting the mirrors of existence

Polarity: feminine; Crone, the sacred marriage of opposites

Gibbous Moon Phase: to improve and perfect

Inanna, the maiden, has no intention of allowing anyone to disturb the relative peace and harmony of her garden, especially because the tree was planted to bring a sense of stability to a chaotic world (the waters). Yet she cannot remain innocent forever, especially within an existence of duality in which the opposite poles of reality will demand to be met and integrated (represented by the inhabitants of the tree).

Without the intricate dance that takes place between opposites such as dark and light and men and women, the hero will fail to achieve the strength and confidence required to become king and will always be afraid of his own shadow.

Thus it is useful to ask:

- If I see myself as strong and capable, how accepting am I of my more vulnerable and sensitive nature?
- If I see myself as generous, where am I a miser?
- If I am overly responsible for others, how easy is it to be irresponsible or take responsibility for and care of myself?
- If I am always good, where do I hold shame about the aspects of myself that I may deem bad?
- If I am always in control, what fear do I have of being out of control?

This sacred marriage of opposites was strongly promoted by the psychologist and alchemist Carl Jung; in his practice he saw the potential damage to the soul's expression when only one pole was given exposure. He stressed the importance of honoring the dual nature of our existence, however uncomfortable, for only then are we capable of finding a level of self-confidence and self-worth not dependent on praise and unaffected by criticism.[8]

Individually, it is easy for us to believe we have created this sacred marriage, while all the time only one pillar of the doorway is being expressed. Hence, it is useful to ask the following questions to reveal any state of imbalance:

- Does this job-relationship-way of life nurture my soul?
- Am I growing spiritually from this situation?
- Am I excited and energized by this experience?
- Do I feel more connected to my soul through this process?

If the answer to any of these questions is no, it is time to make a change. And remember: Just because you're good at doing something doesn't mean you should keep doing it!

Mastery of the Passion

Like Inanna, the awareness of our dual nature is heightened during the first rite of passage of puberty. Here, we not only discover that there is a gender opposite our own, but also we come face to face with the raw and wild part of our nature that expresses itself physically as our body blossoms into adulthood. This expansion of mind and body is represented by three untamable creatures that make their homes within our tree of consciousness:

- The serpent, representing our sexuality
- The bird with the face of a lion and the wings of an eagle, representing our craving for power and knowledge
- Lilith, representing the rebel who will not conform

As we have already learned, Lilith was the first wife of Adam, who refused to lie beneath him, demanding instead that she be treated as an equal. From Hebrew text we learn that when her request is denied, she leaves for the desert and thereafter comes at night to steal other peoples' babies, for apparently she is deprived of her own offspring. Indeed, until quite recently, it was not uncommon for parents to use talismans to keep their babies safe from this dark lady.

Yet the mythological tales of this goddess have become distorted over time, for it is now clear that Lilith is not a sexually frustrated demon but instead a Crone who embodies the purifying fires of the serpent and the clear vision of the eagle. She is represented by the lily and the lotus, symbols associated with the elixir of life, and her message is clear: Through the purification that occurs within our sexual fires and the wisdom that emerges from our hearts, we will come to experience immortality.

To be able to reach this state, however, we must first meet and master these wild energies. Initially, this can appear overwhelming, as every teenager knows. Even as an adult, it can be a challenge to confront the wild and often uncontrollable energies associated with sex, power, and rage, especially if we have been inaccurately taught that spirituality is based on kindness, peace, and love.

In Inanna's world, she finds herself unable to cope with these erratic and erotic feelings and calls upon her brother to destroy the tree and to build a bed and throne from the wood, relegating the essential parts of her creative psyche to the deeper recesses of her mind. She allows his civilizing ax to cut away anything that does not conform to societal norms and to build her a structure where she can fit in.

We see this same pattern over and over again in our own lives; we choose to hide those parts of our personality that we deem unacceptable and project our disapproval and judgment on those who carry such traits. This is perhaps especially true of those in the spiritual movement who advocate love and light through a veil of dispossessed energy. Often, such "fluffy bunnies" are relatively new to the field of spirituality, although I know of some who have reached this place many years ago and have never left. Here, they enjoy the perks of spiritual awakening—spirit guidance, psychic gifts, and synchronicity—while being less interested in true spiritual development, knowing that such development requires personal inner work. When faced with their mirrors, they tend to use pity, forgiveness, understanding, and patronization to deflect the attention. Such individuals may even seek a group who collectively possesses a similar shadow, and together its members can convince each other that denial is the best way to handle any uncomfortable feelings. Any attempt to point out a possible connection between their behavior and their mirrors is met with defensiveness, anger, self-righteousness, and outrage that anybody could possibly be so negative and pessimistic. Then, armed with knowledge of the approaching unified world, they endeavor to avoid conflict (mainly within themselves) by stressing that we are all one big, happy family and anyone who says otherwise is less than spiritual. These are truly the actions of the crab (Cancer), which will shuttle sideways, hide in its shell, or disappear under a rock to evade anything that disturbs the illusion of peace and harmony.

Yet just as it is essential to grasp the head of the serpent—our power—so is it important to turn our attention to our shadow rather than project our dispossessed self onto the world with comments such as: "All the people around me have the same issue; I wonder what is wrong with them!"

It is important to note, however, that while there will always be those who practice such spiritual bypass, the majority hear the wisdom of Athena and turn to face the shadows mirrored in the people and situations surrounding them. We have only to acknowledge our emotional resonance with another person to recognize that we have just met a part of ourselves in this individual. Of course, such emotions do not have to be negative, for we will also resonate with people we admire, respect, and desire to emulate.

Inevitably, those individuals who really get under our skin are often our greatest teachers. Commonly, we find that we have chosen members of our immediate birth family to present us with the greatest reflective surfaces. And if we are still tempted to project our disowned aspects onto the world, I offer a few guidelines, which have helped me to recognize my own mirrors:

- That which we judge in others is what we fear within ourselves.
- When we find ourselves defending our beliefs, we must ask: What is the truth and what is an illusion?
- When we project our shadow onto others, it will inevitably be returned multifold.
- When our reaction to an event is excessive to the situation, we must ask ourselves why.
- What we don't own we will often marry!

As the hero strives to become king by bringing together the opposite poles of existence within a sacred marriage, it is important to remain focused on the goal and objective, especially when reactive feelings arise. If we can remember that even in the most difficult situations, we are still the creator of our perception of reality, we can watch as the energy diminishes once we start to own our shadow. In other words, when we choose not to become a victim to our choices, life becomes much easier.

At the same time, by allowing our Lilith to return from the desert, we are being given the opportunity to release the fear and shame that so many of us have carried with regard to our deeper passions. This allows these fires to burn with the brightness of creative endeavor. As this occurs, we find ourselves embodying a confidence so powerful that it can challenge any man or woman who has not yet mastered his or her own creative energies; we become someone who knows who he or she is, who is comfortable with the passion of sexuality, and who has the eyes of an eagle piercing through the veils of pretence or deception to reveal the essential nature of the soul.

CROWNING THE KING-QUEEN: SUCCEEDING IN THE QUEST

The crowning glory of the hero occurs when his dreams manifest into reality and, as a self-actualized man, he becomes sovereign of his own life. In mythology, the hero-king seizes the sword, wins the fair maiden, makes peace with his father, and takes the throne. In our modern world, the prizes are slightly different and include:

- Graduating from college or receiving a master's degree or Ph.D.
- Owning our first home
- Getting married
- Giving birth to our first child and taking on the mantle of parenthood
- Returning a healthy profit in our own business
- Publishing a first book

Whatever the success, the feelings are the same: elation, celebration, and pride. Yet for some, the moment of glory is diminished by self-deprecating beliefs that deny self-congratulation:

- Nobody likes a big head
- Think of others, not just yourself
- It is wrong to be full of pride
- Someone else did better

Such sentiments are extremely detrimental to the creative cycle, which is dependent on the king's ego being well developed and unaffected by both criticism and praise. In essence, it is the strength of the ego that provides the energy for the next stages of the journey and without which the descent cannot take place. The perfect archetype of such energy is seen in Leo the lion, king of the jungle. He gains respect from others not only due to his physical strength and mighty roar, but from the way he strolls through the jungle with a self-assured air, knowing who he is and proud of his achievements.

♌ LEO

Quality: fixed; self-individuation, self-actualization

Alchemy: conjunction; the king is crowned

Polarity: feminine; king, success, pride, and celebration

Full Moon Phase: conscious integration and success

It is in honor of the Great Mother that we stand completely present in the full bloom of our achievements. As alchemical vessels, we have accomplished a miracle, transforming a dream into reality and bringing the spiritual blueprint to life. This is a moment to celebrate, for often simple gratitude is not enough; it can mask the belief that "I don't deserve this." In my life, I embody all successes with celebration and in those moments I hear the universe rejoice.

As king-queen, then, we must ask:

- What have I successfully brought into reality or manifestation from seed?
- What qualities am I proud to own?
- What do I celebrate as a soul?

It is a wonderful experience for families to share their celebrations, recognizing that this is one of the richest offerings we can give each other. When we hold ourselves in a healthy state of self-confidence, we encourage others to do the same and thereby pass this gift to subsequent generations. The following story reminds us how easy it is for an ancestral message to become distorted in its telling until its truth is almost lost.

> There was a beautiful old priory where ancient religious scripts were held and updated from time to time to accommodate the language of the day. The scribe chosen as translator was greatly honored to be picked because he was selected from a distinguished group of keen and devout priests.
>
> Descending into the vault, he would work all day, ascending from the musty tomb only to sleep. One day, the scribe failed to return to the surface, causing much concern among his colleagues. As they made their way down the cold, stone steps, they heard crying coming from the vault below.
>
> Moving quickly, they found the distinguished scholar in tears, surrounded by many aged books. "What's wrong?" they asked in dismay. Through his tears the priest replied: "He didn't say celibate, he said celebrate!"

Now with the first length of the magician's wand complete, the ida, the hero turned king, has every right to be pleased with himself and, like the mighty lion, can share his celebration with pride. Success is complete—but fulfillment is another story, as the following chapters will reveal.

7

AND THEN ONE DAY . . .

As queen of Sumeria, Inanna has everything: two children, a husband, people to rule, and wealth beyond her dreams, and yet one day she hears the call from the great below, the underworld, and knows it is time.[1] She abandons her temples and responsibilities and prepares for her descent, gathering together seven of her most precious ornaments of status to take with her. She also instructs her faithful servant Ninshubur that if her mistress doesn't return, she should ask for the help of first the great god Enlil, then her father, Nanna, and finally the wise god Enki. Inanna then begins her journey of descent into the darkness, which is the home of her sister, Ereshkigal.

When she reaches the outer gates, she is met by Neti, the gatekeeper, who asks why she is on a road from which no traveler returns. Inanna replies that she has heard that Ereshkigal's husband, Gugalanna, the Bull of Heaven, has died and she wishes to witness the funeral rites. Hearing this, Ereshkigal, the dark sister, allows her to enter but commands Neti to take a piece of Inanna's royal garment from her at each of the seven gates or portals that Inanna will encounter during the descent. "Let the holy priestess of heaven bow low," demands Ereshkigal.

And so starts Inanna's descent into the underworld.

So what is the impetus that causes Inanna to abandon her comfortable life and begin her journey into the underworld? Her intuition; reminding her that the path to spiritual fulfillment cannot be measured merely by

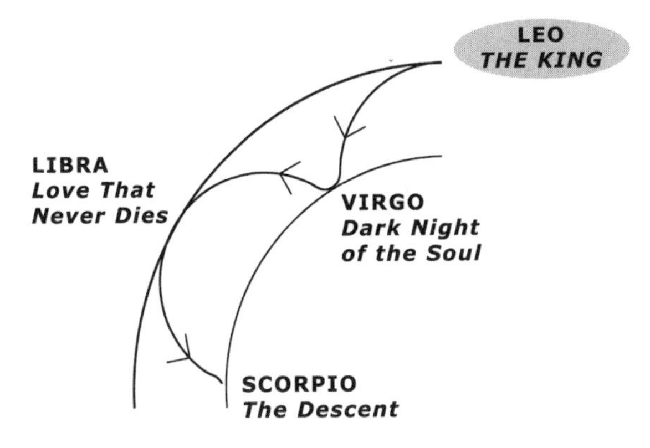

The three levels of descent

earthly success. This highlights the symbolism of the sign of Virgo where Virgin and Crone hold open the portal to the underworld, emphasizing the importance of accepting all parts of the self into our hearts.

♍ VIRGO

Quality: mutable; blending life and form, introspection and contemplation

Alchemy: putrefaction and fermentation

Polarity: feminine; Virgin aspect of the Triple Goddess, the vesica piscis

Disseminating Gibbous Moon Phase: to distribute and convey

This is a good moment to remember that the purpose for our existence is not merely to return to the Great Mother, but also to bring with us the result of the manifestation of our dreams, held as gems of wisdom within our stories. Thus the next three phases of the journey represent the movement of energy from the crown chakra to the base as we eat through and absorb the flesh of our experiences and so create the second length of our magician's wand, the *pingala*. This destructive force, in opposition to the creative power that saw the boy crowned king, brings about the necessary tension for the nurturance of the third energy, the sushumna, and the final extraction of the elixir of life.

When we add the five aspects of the journey from Aries to Leo to the three that culminate in the creation of the pingala, we create the vibration of eight, the number of a new octave, recognizing that everything is in perfect transformation.

Each of us will take at least one journey into the underworld during this lifetime, and indeed, at present humanity as a whole is held tight within the clutches of the Dark Goddess who demands transformation. As in Inanna, the call to descend arises from deep within our soul, entreating us to look beyond the glamour and success of the king-queen and enter the domain of the Triple Goddess to fulfill our true spiritual destiny. Yet as we were taught in Samudra Manthan, focused concentration and willpower are not sufficient to sustain our quest; we must equip ourselves with the energy of the turtle, turning our senses inward by practicing contemplation and introspection.

These qualities of the turtle arise from the fact that in many cultures the turtle is seen as living within two worlds. Hence, it is known as the keeper of the door: it swims with grace through shallow waters and comes onto the land to sleep and lay its eggs, using its highly developed intuitive senses to move between the worlds. In many cultures the turtle is associated with longevity, the primal mother, lunar cycles, and feminine energy and hence, as we will see later, has strong ties to the sacred geometry of the vesica piscis, the doorway that the Triple Goddess provides for us to enter into new realms of existence.

When the call comes, it may appear outwardly that there is little reason for discontent, and yet inside, we often feel dead, empty, or sad. Sometimes, this void appears when our children leave home or when we reach the goal of success but we lack the sense of true fulfillment.

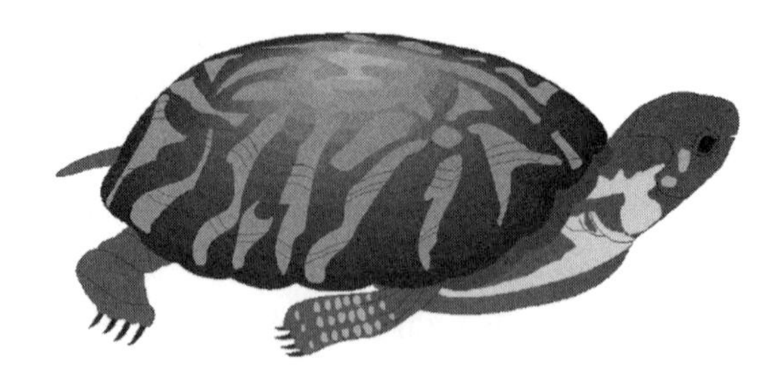

The turtle, the primal mother

At other times, we are pushed into the descent by a crisis such as a job loss, death of a loved one, illness, or divorce. Frequently, however, this dark night of the soul arises from nowhere, appearing to an outsider as if it has arrived as a bolt out of blue, and yet there is no question that we will not accept its invitation.

If we fail to listen, our higher self, the Virgin, is at hand, reminding us that it is time to keep our promise to the Great Mother. To reinforce the call for change, our body and mind express the disharmony, causing us to feel:

- Overwhelmed, stretched too thin, and burned out
- Lost, confused, and disorientated
- Disconnected from what matters in our life
- Frustrated, irritated, angry, and resentful
- Unheard and unseen
- Sad, depressed, and dispirited
- Detached and alone

Physically, it is not uncommon to feel like a coiled spring that cannot take any more. This may be expressed as pain in the muscles and joints, headaches, indigestion, temperature imbalances, palpitations, panic attacks, and severe mood swings. Many modern diseases, including depression, fibromyalgia, chronic fatigue syndrome, endocrine imbalances, and other immune-based illnesses reflect a separation from our soul's path as well as a reluctance to let go and trust the inner wisdom of the Virgin.

Of course, any of these signals, including illness, can be interpreted as inconvenient irritants rather than intuitive lifelines. With free will always an option, we may choose to lessen their interference through drugs, alcohol, positive affirmations, or by losing ourselves even further in the busyness of the outer world.

Yet Pallas Athena is in no rush, living as she does in the place of timelessness and amused by our stubborn belief that we are still in control of our lives. One day, we wake up to the fact that, despite all the tricks in our bag, we are lost, and we cry out to her: "I am ready. Close all the doors that are not in harmony with my soul and leave open or open those that are."

In the end, it is our ability to surrender to the radiant embrace of the Virgin, the expression of the sign of Virgo, that provides freedom from fear, for we know that on some level our journey is already complete.

THE DARK NIGHT OF THE SOUL

Once we are willing to accept our fate, like Inanna, we can prepare for the journey, withdrawing from situations and people who take too much of our energy and even seeking a sabbatical from work. As we collect our few precious belongings for the journey, we celebrate our achievements, knowing that the strength to descend is based on the strength of the king-queen's self-esteem. We may also find that we choose to share our decision only with those who will understand and, in particular, with those who, at some time in their life, have themselves passed through to the other side of a dark night of their soul.

It may not always be friends and family who can offer support during our descent, for they are linked intricately to the present status quo and often prefer that we do not rock the boat. Hence, it is common for outsiders such as therapists, pastors, and other health care professionals to offer a detached and yet compassionate approach to our decision. Anybody who agrees to become the sacred vessel for someone else's alchemical transformation must respect the nature of this calling, for while the heart and not the head is directing the show, the traditional tools of fixing, rescuing, healing, and even analysis may be inappropriate. Indeed, they may be of more service to the practitioner than the client, especially when the former still has concerns about the realm of the Dark Goddess and the chaos and destruction she spawns.

I find that I like to honor those who have, with wisdom and love, created a sacred space for me during my numerous trips into Ereshkigal's domain. I have even found that pets have an uncanny way of sensing the moment we decide to turn within, offering unconditional love that never wanes. There is, of course, one aspect of our nature that also never leaves us: our intuitive self, represented by Inanna's Ninshubar, who knows the importance of the journey and promises to hold the light as we descend.

THE REASON TO DESCEND

As Inanna stands at the outer gate, she offers the excuse that she has come to witness the burial rites of her brother-in-law, the Bull of Heaven.

Like many of us, Inanna would prefer to be the witness of someone else's metaphorical death rather than experience it herself. We can read about it, act it out in shamanic rituals, and even be present at someone else's dark night of the soul, but spiritual prosperity is forthcoming only when we are willing to experience the descent for ourselves.

Psychologically, the death of the bull symbolizes the death of our personal desires and the realization that true wealth or the golden light of pure consciousness is far more precious than anything acquired during the Taurean stage of the journey. Even though we climbed the ladder from the base to the crown chakra, there is still something missing, and our king-queen is determined to turn and descend into the darkness of the unknown to reconnect to the gems of our existence.

Ereshkigal, the Crone, is unsure of her heavenly sister's intentions to visit but allows Inanna to enter on the condition that she "bow low." This is a reminder that humility and surrender are essential at this stage of the journey, releasing the need to:

- Stay in control
- Understand what is happening

- Never feel pain
- Receive prior warning of unpleasant events
- Determine both the pace and the outcome of the experience

This is the realm of the deep unconscious, and it is only through our heart's intuitive pulse that we are reassured that we are exactly where we need to be.

THE ONE: THE MYSTICAL YONI (VULVA)

As we, like Inanna, face the portals through which we must pass in order to reach Ereshkigal's domain, we find each is shaped as an oval, reminding us that we are in the presence of the Triple Goddess, otherwise known as the One, derived from the word *yoni* or *vulva*. To ancient peoples the vulva was sacred, and whether we emerged from it as a newborn baby or entered into it during the act of sex, it was recognized as an opening or portal into another dimension. In mythological terms, our masculine aspect leaves the vulva of the Goddess as the puer and returns as a lover, to embody all that he has achieved.

Mystically, it was believed that life would never be the same once we passed over this magical threshold, but as the honoring of the feminine faded, so did our understanding of the role of the Triple Goddess in the sacred act of sexual intercourse. We forgot that it is the Virgin who excites us sensually, the Crone whose ascending fire burns away any limitations, and the Mother whose waves we play in blissfully at the height of our ecstasy. Men and women alike can provide the feminine vessel both for themselves and a partner, whether the orgasmic release is related to the sexual act or is a natural expression of spiritual reconnection.

During the past millennium, the mystical qualities of the yoni, or vulva, were corrupted and almost lost, obscuring the door to our spiritual heritage. The Triple Goddess, however, heard our cry and has returned to reclaim her children with a resurgence of interest in all things feminine, exemplified by the runaway success of Eve Ensler's *The Vagina Monologues.*[2]

The Vesica Piscis

The most common depiction of the yoni is the vesica piscis, formed when the circumference of one circle passes through the center of another, creating between them a two-pointed oval known as the portal into a new state of existence. (See the illustration on page 35.) The circles themselves are seen to represent opposite poles of existence that come together while maintaining their own unique identity: Unity through diversity. The willingness of the circles to unite in such an intimate way symbolizes the love and respect we must have for ourselves and for our spiritual blueprint before we can attempt the descent. Without such a healthy sense of self or self-esteem, we will cling inevitably to any external source of identity, causing ourselves great pain and angst as the Dark Goddess seeks to reveal our inner core.

As the circles come together to create the third inner shape, we see reflected the three aspects of the Triple Goddess: Creator, Nurturer, and Destroyer. Together, they give meaning to the saying: "When two or more are gathered together . . . there am I in the midst of them" (Matthew 18:20).

This simple message alludes to far more than the physical presence of a single person. It reveals that if we wish to know the oneness of eternal life, Christ consciousness, then we must embody all three faces of the Triple Goddess, which includes the acceptance of the cycles of death and rebirth. At the same time, the complex nature of the forces involved in this sacred geometric pattern remind us that it is only through the harmonic interplay of opposition and attraction, light and dark, that the doorway to our multidimensional existence is available to us.

As we turn away from the external source of success and enter the stillness of introspection and contemplation, we create for ourselves our own vesica piscis, passing through its center to become the lover whose strength and compassion carries us forward toward our spiritual destiny.

Gothic cathedral entrance

THE GODDESS IN SACRED ARCHITECTURE

As we continue our exploration of the Triple Goddess, it is fascinating to note that around the world there are many architectural representations of the yoni, although its identity has often been lost along with its significance as the doorway or entrance into Christ consciousness.

From the twelfth through the fifteenth centuries, reverence of the Goddess was kept alive in the designs of many of the great Gothic cathedrals of Europe, which were built with magnificent oval arched doorways, symbolizing the vesica piscis, leading into the nave, or womb, of the church. Looking back further, certain Greek rectangular temples were built around the geometry of the vesica piscis: the dimensions of the side walls of the temples were based on the ratio of 1 to the square root of 3, to heighten the reverence within.

Perhaps the architects of these buildings wanted future generations to know that

- It is through the feminine within each of us that we can reach the unified field of the kingdom of heaven.
- The unity symbolized by Christ consciousness is accessed through the power of love and leads to immense feelings of joy.
- The sacred geometry of these buildings emanates a frequency that allows us to meet our God-Creator eye-to-eye, without the need for an intermediary.
- All—no matter their creed, culture, or gender—are welcome in this place as long as they are open-hearted and joyful.

It is obvious that the present-day practices of certain religions have strayed far from the conditions laid down by these designers. At the same time, many of today's architects still do not understand how the vibrational qualities of sacred symbols influence the overall function of a building. Even when there is an acknowledgement of the power of shapes to enhance our experience within a structure, many modern buildings have no curves, instead employing straight lines rising skyward, leaving little doubt as to who is boss.

Sheela-na-gigs

In Ireland between the thirteenth and seventeenth centuries, architects went even further to ensure that the congregation appreciated that it was through the Great Mother that they would come to know their God: These builders created carvings called Sheela-na-gigs. These statues, often found over the doorway of a church, depicted naked females with a full belly, posing in such a manner as to display and emphasize the genitalia. Many of them revealed a woman with her knees apart and the vesica piscis–shaped hole of the vulva held open by one or both of her hands.

There are various accounts of the symbolic meaning of these images, including a particularly patriarchal view that they were produced to protect men from eternal damnation by reminding them of the lustful nature of women. It is more likely, however, that they

Sheela-na-gig

represented something far more significant than a by-product of Christian dogma. The meaning of Sheela-na-gig is best translated as "vulva woman."

These figures closely resemble the yonic statues of Kali, which appear at the doorways to many Hindu temples to confer good luck on all who enter. The protruding rib cage seen in several of the Irish carvings is also prominent in statues of Kalika, the death goddess of the Hindus. She is symbolized in Irish tradition by Caillech, or the Crone or Hag, who is both the creator and the destroyer. Thus the Sheela-na-gig is seen to represent all three aspects of the Goddess—Virgin (yoni), Mother (belly), and Crone (ribs).

There can be little doubt that the Celtic Christian church in the first millennium was far more liberal and aware of the importance of the feminine within its society than its Roman counterpart—for example, it allowed divorce well into the twelfth century. Eventually, the ancient Celtic contribution was suppressed and the Roman Catholic influence promoted these effigies as crude, exhibitionist images. Yet, we might prophesize that the Sheela-na-gigs will soon be returned to their rightful places at entrances to the Mother Church as people seek out her womb for direct union with the divine source.

The Triangle

Another symbol commonly used to represent the yoni of the Triple Goddess is the triangle, located at the entry point of many sacred sites and associated closely with the Oriental goddess Cunti, from which are derived words such as *county, country, ken* (to know), *cunning,* and *cunt.* Different from modern-day usage, this last was not a word of derision, but of respect, honoring the embodiment of the Goddess within a woman.

The same derivation gives us the word *kin,* or family, leading to the word *kingdom,* which denotes the domain of a king or queen. In olden days, a kingdom was land that was passed down through a matrilineal pattern, emphasizing the importance of the mother's bloodline. In other words, it was understood that the continuation of a family and its fruitfulness on all levels was dependent on the feminine and her link to Goddess energy.

The Shamrock

There is another symbol found within the Celtic world that denotes the Triple Goddess: a tiny, three-petaled plant called the shamrock. Despite the fact that this national symbol of Ireland is usually associated with St. Patrick and his interpretation of the Trinity as Father, Son, and Holy Ghost, its importance stretches back, far beyond his auspicious arrival on the Emerald Isle. It is now becoming increasingly clear that the "autobiography" of the patron saint of Ireland was written between the ninth and tenth centuries, four hundred years after his apparent ministry. Before the ninth century "pagan" beliefs and sacred practices were openly accepted within a highly integrated religious discipline.

As so often happens when a new paradigm attempts to impose itself onto a much older version, the personal details of actual historical figures are stolen and bastardized to create an entirely new persona that matches the dogma taught at the time. The original character is either sold to the people as a fraud or as a heathen and eventually gets demoted in history as a mythical figure. Thus the pagan representative of Patrick was probably the Irish god of the shamrock, Trefuilngid Tre-eochair, who was the son and consort of the Triple Goddess of the land and whose sacred plant bore edible fruits, including the apples of immortality. It now becomes obvious that the shamrock reflects the triple yoni of the Great Goddess, a symbol dating back to about 2500 BCE. With this in mind, a whole new meaning is brought to the wearing of the shamrock on St. Patrick's Day.

The Celtic Cross

The shamrock was probably the model for yet another important Celtic symbol reflecting the sun wheel, later modified into the present-day Celtic cross. With its three short arms and long vertical arm, the Celtic cross captures the image of the phallus held within the triple-faced yoni—the marriage of the male and female, the symbol of fertility and the continuation of life.

Indeed, the design of the most sacred monument in Ireland, Newgrange, is based on the principle of the Celtic cross. Walking down a long passageway (the phallus), visitors enter a tall, round cavern from which three smaller openings (the Triple Goddess) emerge. For five days around the winter solstice, if the weather is clement, the rays of the sun enter a small box above the entrance and pass down the corridor, eventually penetrating the cavern, filling it with light.

To the mystic, this act symbolizes the union between the dying king and the Great Goddess, who gives birth to a whole new frequency of consciousness as the sun returns to the northern hemisphere. From the depths of Newgrange, the Mother nurtures the new baby until she deems it is ready to leave the confines of its stone home. From there its unique message will be spread throughout the globe via the complex

The Celtic cross

grid system that exists within the earth, so that all living things can be touched by its energy.

Around the world there are many sacred sites aligned to the solstices. Those, like Newgrange, linked to the winter solstice, highlight the Virgin's love in giving birth to and nurturing the puer until he leaves home as the hero. On the other hand, those linked to the summer solstice, such as Stonehenge, are connected to the dying sun and represent the beginning of the descent of the king into the underworld, where, six months later, he will eventually offer his life for the sake of his son and heir.

The Spiral

In front of the entrance to Newgrange is a large oblong stone with intricate carvings displaying the three-spiraled head of the Celtic cross, reminding the visitor that this is the sacred domain of the Triple Goddess. The symbol of the spiral is found throughout the world, carved into the stone of many sacred sites. In most cultures, the glyph depicts the great serpent mother, who is said to have created the entire world

Newgrange, Ireland

from the place of formlessness. To the followers of Kali, she is known as kundalini, whose curled up and unmanifested self is represented by a dot. The unfolding of this creative serpentine energy into form and its return into formlessness is symbolized by the spiral, reminding the observer that the movement between form and formlessness is two-directional and continuous and the principle behind immortality.

The spiral is seen to be the most effective way of enhancing creative energy. The fact that the spirals found on the walls of the inner caverns of Newgrange are aligned to the lunar standstills reminds us of the powerful influence of the moon on our own creative cycle: These astronomical events are associated with maximum fertility for mind, body, and spirit.

The Labyrinth

Before we fully embark with Inanna on our descent into the underworld, there is one more piece of equipment we require. This interactive map has been used by most indigenous people around the world and is known today as a labyrinth. As a unicursal or one-path maze, it derives its name from the Greek word *labrys,* which means "double-headed ax." Most traditions, however, see the labyrinth as a much older symbol, representing the sacred womb of Mother Earth, the goddess waiting patiently at its center. Traditionally, labyrinths were guarded by women, for only they were acknowledged to understand the principles behind the cycles of death and rebirth.

Some of the earliest examples of a labyrinth date back to around 5000 BCE and were built in Mesopotamia in honor of Inanna. The Hopi Indians also see the labyrinth as an entry point to the underworld and the source of rebirth. They liken it to their *kivas,* or underground sanctuaries, from which, it is believed, all Hopi people originally emerged. In southern India, Hindu women still mark their homes with the sign of the labyrinth at the beginning of the decline of the sun at the summer solstice.

The classic design of a labyrinth has seven circuits or paths leading to the center. This is often called a Cretan labyrinth, although the

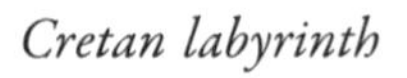

Cretan labyrinth

labyrinth dates from a much earlier civilization. As the seeker soon discovers, these circuits are not in sequential order, which encourages the traveler to surrender the desire to control the outcome. At the same time, this symbolic shape brings emotions and beliefs to the surface; these must be cleared before we can stand "naked" before the Great Goddess. The seven paths of the labyrinth represent the seven chakras or portals of consciousness through which we must pass, although, contrary to common protocol, we descend from the crown chakra to the base chakra as we enter the labyrinth and ascend back to the crown on our return to the outside.

Although many labyrinths today are built on flat land, there are a few built either on the top of a hill or into a hill, the most famous being the three-dimensional labyrinth that winds around Glastonbury Tor. As the pilgrim walks this seven-path labyrinth, he or she is once again given the opportunity to offer to the Great Goddesses those things that have become unnecessary baggage and as such are hampering the deeper journey into the self.

Mythologically, the best-known tale concerning the labyrinth is that of the Minoan hero Theseus, who enters the labyrinth with the aid of Ariadne's thread to meet and kill the Minotaur, who is half man and half bull. The story, however, often fails to include the fact that Ariadne was a powerful moon goddess and that the Minotaur was Theseus's own inner demon. The story ends with Theseus abandoning Ariadne after achieving his goal, although we can imagine that this was not his wisest move!

Many centuries later, an eleven-circuit labyrinth was designed and

Chartres labyrinth

first appeared in Chartres cathedral in France. It was thought that the labyrinth was dedicated to the Virgin Mary. Called the Road to Jerusalem, it was generally considered that by walking this path, often on his or her knees, the pilgrim could repent his or her sins and hence become closer to God. Whether it was appreciated that this process involved entering the womb of the Dark Goddess we will never know—but the cathedral was built on top of a hill where several ley or earth energy lines intersect, leaving no doubt, perhaps, that its architects, the Knights Templar, knew that this was a perfect portal for entry into the underworld. Here, the pilgrim would meet not only the Virgin, but also the Crone, personified by the Black Madonna—thought by some to be Mary Magdalene—and would come to experience the true meaning of death and rebirth.

This theory is reinforced by the intricate design of this beautiful labyrinth, which consists of four quadrants and, in its center, a six-petaled rose. The four quadrants signify the four aspects of form or structure that reflect the development of the ego until the time when the king reigns supreme. These include the four directions, four elements, and the four functions of the mind. The six petals symbolize the seed of life and the promise of rebirth. Only by walking and "unwinding" the life we have created can we enter the void, the center, and receive the seed of new life.

So let us be clear before we proceed:

- The only place we should want to be at present is here on planet earth.
- The most important vehicle for the journey is the physical body.
- The most powerful energies available for transformation are the sexual fires.
- The only place that we can achieve alchemical transformation is in the underworld, otherwise known as hell.
- The unconquerable force is love, which will persist until we remember all aspects of ourselves and become whole. That is the power of the Great Mother.

The great deception has managed to keep these facts hidden from us until now!

8

THE LOVE THAT KNOWS NO END

It may seem strange that the phase of the journey related to love is masculine, associated as it with the sign of Libra, the symbol of balance. Yet the love expressed in the following stories is dynamic and carries all the hallmarks of courageous acts that are symbolically associated with heroism. Indeed, it takes a deep self-love to choose to face those parts of the self that have become separated and which, even now, are languishing in the darkness. Those of us who have experienced the torment or sadness of loss know that it is only through a reconnection, either physical or spiritual, that we can truly come to know the meaning of sublime love.

♎ LIBRA

Quality: cardinal; conscious appreciation of the creative plan

Alchemy: putrefaction and fermentation

Polarity: masculine; lover, compassion, courage, accountability, self-discipline

Disseminating Gibbous Moon Phase: to distribute and convey

It is thus the lover within us all that can relate to this archetype whose compassion, as we learn here, knows no end.

DEMETER: THE GREAT MOTHER

One of the best known mythological characters associated with love is the goddess Demeter, whose grief for the loss of her daughter Kore and her subsequent trials to ensure that her daughter is returned safely are told in detail in the *Homeric Hymn to Demeter,* thought to have been written around the seventh century BCE. Many readers may have heard that Persephone was Demeter's daughter. Yet since Persephone is a Triple Goddess, the name of the Virgin aspect is Kore; the Crone aspect is known as Persephone.

> Zeus, ruler of all gods and men, promises Kore (his daughter from a brief marriage to his sister Demeter) secretly to his brother Hades, god of the underworld. After the plot is agreed upon, it is just a matter of luring Kore from the watchful eye of her mother. This occurs one day when the maiden's attention is caught by a particularly beautiful six-petaled flower, the narcissus.
>
> As she bends to pick this object of fascination, the ground opens and she is seized by Hades (also known as Pluto) and taken, crying, into the underworld aboard his golden chariot. According to the Greek version, we hear little more of Kore's plight until one year later, although the Roman version adds the suggestion that she was raped, in accordance with the dominant influence of the masculine energy at that time.
>
> Meanwhile, on the surface, Demeter is beside herself with worry, entreating immortals and mortals alike to share what they know—and yet all are silent. Even though Helios, the sun god, and Hecate, the moon goddess, heard Kore's cries, they say nothing to her mother. For nine days Demeter flies around the world, refusing food, drink, and sleep until her daughter is returned. On the tenth day, Hecate comes forward and tells of the plotting between the brothers, and how Hades plans to make Kore his wife. Naturally, Demeter is livid and demands action from Zeus, who merely turns his back on her.

In revenge, she immediately withdraws all the nurturing energy from the earth and begins to wander aimlessly among mortals, concealing her goddess powers and appearing as an old woman. It is said that another nine days pass (interestingly, nineteen in all; we will remember that nineteen represents the acceptance of death and rebirth), before she meets the four daughters of Keleos, the king of Eleusis, beside the city's well. When they hear that she is looking for work as a nursemaid, they are delighted, and she is admitted to the royal household to take care of their infant brother, Demophon.

On first sight, the queen, Metaneira, knows that there is something noble about this old woman, especially because the shadow Demeter casts in the doorway is radiant. Yet Demeter refuses the comfort of a splendid chair and the sustenance of wine, preferring to rest on a rustic stool and partake of a simple barley drink.

Now a nursemaid to the royal child, Demeter secretly determines to extend the gift of immortality to him, having lost her own child. She feeds him the ambrosia of the gods (the same as the Hindu amrita), breathes her sweet breath upon him, and at night holds him over the fire to bring out the best in him. His parents marvel at how well he looks and yet are unaware of the reason, until one night the queen walks in on Demeter and sees the royal heir smoldering within the flames of the fire. Shrieking, she demands an explanation, and Demeter reveals herself as the golden-headed goddess that she is. She then admonishes the queen for preventing her son from becoming immortal and declares that from this day forward, every year, the sons of Eleusis will be involved in a terrible battle. Yet she comforts the queen by saying, "Build me a temple and I will teach you the secrets and sacred rituals of immortality, which will ease your pain."

Thus the Eleusinian mysteries were born (in approximately 1500 BCE). They took place annually for a period of almost two thousand years, until the Roman authorities closed them down. Initially, admission to the mysteries was allowed only to those who had not shed blood and who were Greek, but over time some of the restrictions were lifted, allowing thousands to flock to the little town of Eleusis every year.

Despite the numbers, however, extreme secrecy surrounded the proceedings except for the preparation. What is clear is that during nine days in September, much of the time was spent reenacting Demeter's story. The climax was the lighting of a fire that symbolized the presence of life after death. Today, the town of Eleusis is one of the most polluted areas of the Mediterranean, full of large oil refineries. It is a far cry from its glory days of the Great Goddess.

Meanwhile, back in ancient times:

> The land has been barren for almost a year, and despite various entreaties, Demeter still refuses to restore the earth's vitality until her daughter is returned. Zeus sends to her all manner of envoys laden with gifts, but she refuses to relent. Finally, he concedes to assist and sends Hermes, the messenger of the gods, into the underworld to bring back Kore, although by now she has become Persephone, queen of the underworld.
>
> Hades, unwilling to give up his consort so easily, offers Persephone seeds from the pomegranate to sustain her on the journey back to the surface. Eager to return home and despite her mother's previous warnings to avoid any food from the underworld lest she become entrapped, Persephone eats the seeds. On the surface once again, Demeter is delighted to see her daughter but realizes Hades did not let her go without conditions. She knows that having eaten the seeds, her daughter is destined always to return to the underworld for at least one third of the year, signified by the darkness of winter. During this darkness, the Great Mother withdraws her energy from the earth, replaying her grief for her lost daughter until she can once again rejoice at her return, at which time she showers the earth with spring's abundance of color and beauty.

Despite a belief that the story of Kore is a battle between gods and goddesses, the tale introduces many similarities between Kore's descent into the underworld and the journey undertaken by both Inanna the Sumerian goddess and Ishtar, her Babylonian equivalent. Both Inanna and Ishtar descend to meet their dark sister, which, when we consider Persephone as the Crone, is probably the purpose behind's Kore's journey.

This explanation becomes more probable when it is revealed that the masculine energies of the gods Pluto and Hades were added into the story at a later date, for the underworld was originally the domain of the Dark Goddess or Crone alone. Thus it was Kore's destiny, as it is ours, to descend into the darkness, transforming the ethereal innocence of our external experiences into the deep wisdom of essential knowing.

INNOCENCE
The manifested form of our spiritual blueprint
. . . gives way to
IN NO SENSE
"out of one's mind and out of control"
. . . gives way to
INNER SENSE
The wisdom of conscious knowing distilled from any experience
= knowledge in action

Apart from Helios, it is only Hecate, the goddess of crossroads, who observes Kore's passage. Seen commonly with three heads, she reflects the ability to move with ease between the worlds. She is our silent witness who knows that despite the suffering that comes from leaving behind what is familiar and comfortable, the time will come when we too must descend to meet our dark sister and be transformed in her presence.

Kore and Narcissus

In Greek mythology Narcissus is a young man who rejects the romantic advances of the nymph Echo and is punished by being doomed to fall in love with his own reflection as seen in a pool of water. Unable to consummate his love, he eventually pines away and is turned into the beautiful six-petaled flower that bears his name. From this myth we extract the psychological term *narcissism,* implying someone who is obsessed with self-love.

In Kore's tale, we glean a different meaning of the word, especially when we understand that the six petals symbolize the potentiality of new life. It is her self-absorption and an instinctual desire to move away from superficiality and innocence (the reflective pool) that causes her to reach out to pick the flower, which seals her fate immediately. It is clear that a certain amount of self-love is very important for the psyche and no more important than during this phase of the journey. Here, we listen intuitively to the heart of the lover and sense a deep yearning to reconnect to the parts of the self that have become separated, even though doing this requires facing the unknown depths of our psyche. As we gain the courage to push away the boulders we ourselves have placed across the entrance to the underworld, our life takes on a richness previously experienced only in our dreams.

If, like the youth Narcissus, we deny the enticements of our inner calling—the nymph Echo—those deep feelings will rise up to meet us, bringing chaos, confusion, and anger and forcing us to release our hold on a romanticized state of spirituality and dive fully into the cauldron of transformation.

Demeter's Grief

Most of the *Homeric Hymn to Demeter* is concerned with Demeter's anguish as she passes through the classical stages of the grief process:

- **Numbness and denial,** where she is denied answers
- **Anger** at Zeus for tricking her

- **Withdrawal** as she removes the nurturance to the land
- **Blame of self** as she wanders, denying her own nature and need for nurturance
- **Bargaining** as she offers the gift of immortality to another child in return for his allegiance to her
- **Confusion, disorientation, and rage** when her plans are stopped and she has to face the reality of the loss of her daughter
- **Acceptance and the willingness to move forward again** with the building of the temple and the beginning of the Eleusinian mysteries based on her grief

Demeter's story actually records the seven doorways through which we all must pass during any descent into the darkness. At each door we surrender the masks, veils, and regalia of our outer world until we rediscover our true inner nature.

The Regenerative Powers of the Pomegranate

This red, hard-cased fruit whose name means "apple seeded" is packed with seeds that contain a blood-red juice. Similar in design to an ovary with eggs within it, the pomegranate is seen to represent the cycle of death and rebirth that is essential for an eternal life. Because of the relation between the fruit's seeds and the ova, Persephone's decision to eat the fruit opens a passageway between the world below and the world above, offering all the freedom to move between both worlds. At the

The pomegranate

same time, she determines that all women will follow the patterns of the moon, spending the majority of the time above the earth in the light and less time, that of the monthly menses, within the darkness of the underworld.

An Alternate Version of Kore's Story

There is another version of the Demeter-Kore (Persephone) tale, Charlene Spretnak's version, which sheds a slightly different light on the subject.[1] This Cretan version also describes Demeter as the goddess of the grain, but this time her daughter is seen as her helper, gathering wildflowers and, in particular, red poppies.

> One day, Kore approaches her mother and asks: "Who takes care of the spirits of the dead that drift around their old homes and families aimlessly?" Her mother admits that it is her job to look after the underworld, but she explains that she has little extra time, working as she does to maintain the fertility of the land. Undeterred and unafraid, Kore volunteers to take on the job alone.
>
> Her mother is unhappy about her daughter's journey into the place of the dead and weeps as she gives her a torch to light the way. Kore gathers red poppies, pomegranates, and sheaths of grain as offerings from the earth goddess and descends into the darkness. She wanders until she finds a large cavern where the spirits of the dead live. Here she sits, placing before her a dish of pomegranate seeds, the food of the dead. She then calls each spirit in turn to come before her so that she can embrace him or her and mark his or her forehead with pomegranate juice, the sign of preparation for rebirth into the upper world.
>
> Meanwhile, the land lies barren in the throes of winter while Demeter awaits her daughter's return. Then one day, she sees a crocus pushing its head through the surface of the earth and she becomes ecstatic, for now she knows that spring is returning; the ewes will give birth to their first lambs and soon she will be reunited with her beloved daughter.

This second version reveals Kore not as an innocent child but instead more as a goddess in waiting, prepared to do the work required to take over her mother's role one day. There is no mention of the masculine energies of Zeus or Hades. Instead, the story deals entirely with the cyclical interchange between the three aspects of the Triple Goddess through the stages of death, grief, and resurrection.

Holding forth the gift of rebirth, our own descended Kore gives life to those aspects of the self that have become separated and are, in essence, dead to us. Loosening any old and dying energetic ties that anchor the soul tenaciously to the physical and astral worlds, the juice of life allows the essential light of wisdom to be released from each experience. This essence is then stored in the *ajna,* or third eye, until the "wings" of this energetic center are strong enough to transport our soul between the worlds by its own volition and with ease.

ISIS: THE EGYPTIAN QUEEN OF THE THRONE

When we speak of the theme of love as presented in mythology, our attention turns naturally toward the great Egyptian goddess Isis and her brother-husband Osiris. Isis is a Triple Goddess expressing the power of creation as the goddess of fertility and motherhood and the power of destruction as the goddess of magic and the underworld. As the personification of the throne, her original headdress symbolized an open seat, denoting the fact that all pharaohs had to come and sit in her lap in order to gain the power to rule. This was a prerequisite set by many of the Mother goddesses.

The Grief of Isis over the Death of Osiris

Osiris and Isis rule Egypt harmoniously, bringing justice and civilization to all their people. Yet Osiris has a bitter and jealous brother, Set or Seth, who seeks to take the throne for himself. Through his scheming, he creates a beautifully decorated sarcophagus that is exactly contoured to the frame of Osiris. Set then invites everyone

Isis, the Great Mother

to a magnificent feast during which he offers a gift to anybody who can fit comfortably within the box. Osiris, who is pure of heart, suspects nothing and eagerly steps inside the sarcophagus and lies down. Immediately, the lid is shut and nailed closed and molten lead is poured into the seam to seal his fate. The tomb of Osiris is then thrown into the Nile River, where it awaits finding by his beloved wife, Isis.

In her search, she comes across the box lodged in a tamarisk bush, and she rises from the ground as a hawk, singing the song of mourning. As she soars above the earth, she casts a spell that allows the spirit of the dead Osiris to enter her, and hence she conceives and bears a son, Horus, whose destiny is to avenge his father's death. Scared that the uncle Set may try to kill her son, Isis hides him on an island and returns to Thoth, the lord of knowledge, to obtain the magic necessary to bring her beloved back to life.

Before this can happen, Set discovers Osiris's dead body and, determined to finish his murder once and for all, cuts the body into fourteen pieces, scattering them across the land of Egypt. Yet Isis's

> love does not diminish, and, seeking the help of her sister Nephthys, who has deserted Set, her husband, the two women start to collect the pieces, erecting a temple to Osiris at the location where each piece is found. There is, however, one piece that is never found: his penis.

In some translations, Horus is not conceived until this moment, when a penis crafted out of clay allows Isis to conceive and give birth to her son. All versions agree that with the help of Thoth and Anubis, Osiris is sewn together and his spirit is returned to his body. Because he is dead, however, he cannot return to the living and hence descends to Amenti, the place of the dead, and becomes its lord.

> Meanwhile, Set is about to step forward to claim the throne when Horus returns to avenge his father's death. Together, they enter a bitter feud, which eventually leads to Horus taking the throne while Set is cast into the darkness, where he dwells today, plotting his next move. It is believed that the battle still simmers between them, and that peace occurs when Horus has the upper hand and war and turmoil signify Set's revenge. It is prophesized, however, that Horus will eventually overcome Set, the tombs will open, and the dead will live again, and at that time Osiris will once again walk this earth.

The Alchemical Transformation of Osiris

When we first meet Osiris, he is a much-honored king, and yet, as we now know, he was only half-way through his journey. In alchemical teachings, he must lay down his crown and descend into the underworld to meet his dark brother Set, reflecting a similar journey that is taken by Inanna and, later, her husband, Dumuzi. Seth (Hebrew) or Set (Egyptian) esoterically represents the third aspect of the Egyptian deity triad, in which Osiris is the creator, Horus is the preserver, and Seth/Set is the destroyer or regenerator.

During the descent, (as we will learn later and as stated in the zodiac lists in this and previous chapters), an alchemical process involving *putrefaction* and *fermentation* occurs, in which the meat of our stories and experiences are broken down until only the bones remain. It is therefore interesting to learn that the meaning of the word *sarcophagus* originates from the Greek words *sarx,* meaning "flesh," and *phagein,* meaning "to eat." Hence, Osiris's coffin is flesh-eating, a perfect description of the process.

Once Osiris's fate is sealed, the box is cast into the water, a metaphor for the depths of our psyche where we embrace our own demons. As the putrefaction comes to a close, the sarcophagus becomes wedged in a tamarisk tree, a member of the acacia family with mystical links to the realms of immortality.

Set's final revenge of dismemberment results in Osiris's body being scattered over a number of different sites. Whether there are thirteen or fourteen pieces is debatable. Those who believe there are thirteen relate the pieces to the lunar months of the year, while those who support the idea of fourteen see it as the number of days before the moon waxes or wanes. In either case, the scattered pieces represent those parts of us that have become separated and must be remembered: They are brought back into the fold in the name of love, while the gift each brings to our life is celebrated—in the story, with the building of a temple. This represents the alchemical process of *distillation,* when the essence of each experience is collected and honored, thus building our Ka, or light body, of eternal life.

Finally, we learn that only Osiris's penis cannot be found. Once again, there are two translations. The first suggests that this organ is not required in the world of spirit, while the other believes that it represents physical desire that would bind Osiris to the material world. Either way, without his penis, Osiris's spirit is free to embrace his honored position, the alchemical process of *coagulation*.

Whether at this point or earlier, Isis uses her magic and an earthen penis to become impregnated by her dead husband, eventually giving birth to, Horus, who carries the spirit of the dead Osiris. Thus we are again privy to the sacrifice of the old god so that his son and heir can be born, just as the old sun gives way to the new each night at sunset.

THE POWER OF LOVE

The common theme between this and the story of Demeter is the tremendous capacity of love shown by these powerful Mother archetypes, reminding us that even in our darkest moments, we are never alone. As we descend into the underworld of our own psyche, it may feel as if we are being punished or have been abandoned, and yet it is the Mother's love that urges us on. Just as she grieves when we, her loved ones, die to the old form, so she celebrates our rebirth or return, knowing that it is through her tears that the process of purification will be accelerated.

As a wise Maori healer once told me, "When the pain gets too unbearable, offer it up to the tears of the Great Mother, knowing that her heart is large enough to embrace and transform anything that has become a burden too heavy to carry."

9
THE DESCENT

Hearing that Inanna, her heavenly sister, wants to visit, Ereshkigal allows her to enter but commands Neti to take from Inanna a piece of her royal garment at each of the seven gates encountered during the descent. "Let the holy priestess of heaven bow low," demands Ereshkigal.

Following her command, Neti opens the outer gate and allows Inanna to enter, but only after he has taken the *shugurra,* or crown of the steppe, from her head. Inanna demands to know what is happening, but Neti simply states, "The ways of the underworld are perfect and may not be questioned."

Inanna's descent continues. At each gate another piece of her clothing or an ornament is taken, and each time she asks the same question and receives the same answer.

Hence:

- At the second gate, the small lapis beads from around her neck are removed
- At the third gate, the double strand of beads is removed from her breast
- At the fourth gate, the breastplate called "Come, man, come" is removed from her chest
- At the fifth gate, the golden ring from her wrist is removed

- At the sixth gate, the lapis measuring rod that she holds in her hand is removed
- At the seventh gate, the royal robe is removed

> As she stands naked before Ereshkigal, the latter fastens on her the eye of death, speaks words of wrath to her, utters a cry of guilt, and kills her. Then Inanna's corpse is hung on a hook on the wall as if it is a piece of rotting meat.[1]

Enter the Crone, whose domain is the Dark Rift of the Milky Way, where humanity currently finds itself. There is no avoiding her fiery cauldron if we are committed to passing through to the heart of the Great Mother and merging with the eternal oneness. Yet, as you can see, this is the most challenging of all the phases of the journey, directed as it is by a Dark Goddess, Ereshkigal.

THE CRONE

There is no doubt that, of all the aspects of the Triple Goddess, the Crone is probably the most difficult for an individual to accept and integrate, being both revered and feared within most cultures. Known as the Old Woman, the Wise One, the Dark Mother, and the Hag, she is commonly portrayed as being bloodthirsty, sexually promiscuous, and extremely ugly. The territory she rules is seen as chaotic, representing the unknown and untamed aspects of our nature, which is why those who like to stay in control find her appearance in their lives so disturbing.

Astrologically, this stage of the journey is associated with a scorpion and the sign of Scorpio, which is usually thought to be intense, powerful, dangerous, deep, sexual, and occult—wonderful descriptors of the Crone. Yet for some cultures, her energy is too powerful and intense, causing them to try to erase all knowledge of the Dark Goddess and her association with death from the collective psyche of their people. Hence, we see many people today who would rather ingest chemicals and hormones to stay artificially young than face the Crone and the promise of death and old age. I have also met people who tell me that they are "into transformation" but don't "do death," preferring the immediate (and apparently painless transition) from caterpillar into butterfly while avoiding the cocoon.

♏ SCORPIO

Quality: Fixed; conscious appreciation of the creative plan

Alchemy: Putrefaction and fermentation

Feminine: The Crone aspect of the Triple Goddess; death and chaos

Last or Third Quarter Moon Phase: to revise and re-evaluate

The Crone, however, is not a lady to be ignored, and as the keeper of time, she gives us notice that the psychological defenses of denial and projection will no longer be strong enough to deflect her energies as she rises up to bring about the necessary destruction of the old world to make way for the new.

Already we are seeing her influence in the disharmony and conflict around the world in places where painful issues have been buried and never resolved. Old grievances, betrayals, abuses, and hatred are being forced from their uneasy resting place within the earth and acknowledged and transformed in ways that will lead to lasting peace. The planet itself is helping with this shake up: Witness the increasing number of earthquakes, erupting volcanoes, tsunamis, floods, fires, and hurricanes around the globe.

Now as we face the Galactic Center, the Dark Goddess is demanding that we own our creations and recognize that wherever we experience conflict, fear, or shame, a part of our consciousness is trapped in a story awaiting release. These aspects of the self are now crying out for attention and will not be quiet until we discover the gems of wisdom they hold within their midst.

Kali with the heads of her victims

The Global Crone

To completely understand the descent and, indeed, the ascension through the sage into the heart, it is invaluable to meet some of the representations of this powerful feminine archetype. Known as Kali the destroyer, Cerridwen the corpse-eating sow, Sekhmet the fire-breathing lion, Isis the vulture queen, Morgan Le Fay the death queen, and Persephone the destroyer, it is no wonder she has acquired such a reputation. All represent death, winter, destruction, and doom. Indeed, the Grim Reaper, seen brandishing his scythe as a sign of impending death, originates from an ancient Scythian goddess whose symbol, like that of many of the death goddesses, is the crescent moon.

As we will discover, each cultural myth surrounding the Crone offers a different perspective of her archetypal energy, although all agree that this is one powerful lady!

Sehkmet: The Lion Goddess

It is said that Sekhmet's breath created the hot, desert winds and that arrows of fire darted from her eyes. Originally known as the protector of the pharaohs of Upper Egypt, the Sphinx, which stands on the Giza Plateau, is strongly believed to have been built in honor of this powerful lion goddess as far back as eleven thousand years ago.

There is also a myth that tells us:

> Ra, the sun god, created Sekhmet from his fiery eye and called on her to destroy the mortals who conspired against him. When he saw the blood running in the streets, he knew he must stop her, and hence, he produced a red ochre beer. In her thirst for blood, she drank it, and she became intoxicated immediately, falling asleep and reverting into a benign and loving goddess.

Historically, Sekhmet is probably much older than the sun god Ra, and yet it is possible that the ancient people left us a warning within this traditional tale. It is believed that Sekhmet represents the destructive side of Ra's solar rays, known colloquially as solar flares, and that when there is extreme activity, as we are currently experiencing from the surface of the sun, blood will flow. Over time, the sun will once again fall asleep and return to its benevolent self.

Esoterically, the flow of blood has also been seen to refer to menstrual bleeding, with its release of melatonin, DMT, and other hormones from the pineal gland, leading to an increased psychic sensitivity or intoxication. From this point of view, the story certainly supports Laurence Gardner's research, which tells of the drinking of menstrual blood by the king-shamans to reach heightened states of consciousness before they entered the Great Mother's ocean of abundance.[2] One other theory is that this myth predicts a time when we will collectively pass through a time of apparent chaos and destruction that will lead to our conscious self (the known) falling asleep. When this occurs, our spirit will be free to travel and enter the multidimensional realms of the Great Mother. Eventually, we will calm down and "come down to earth" before a new cycle begins.

Hel, the Goddess of Regeneration

Named after the great goddess Hel, hell is one of the most feared places in religious mythology. Unlike the Christians, however, the ancient Norse people saw the underworld not as a place of retribution and punishment but as a womb of rebirth and regeneration. The earliest shrines to Hel were, in fact, uterine-shaped caverns often connected to an underground hot water stream or steam vent fed by a nearby volcano. At other times, the cave was connected to an ice flow, representing the fact that Hel was comfortable in all extremes of temperature. To the Norse people, everybody, including gods and goddesses, had to pass through the Goddess' domain, and therefore it was not a place to be feared.

In the Pacific region, Mother Death is still believed to live within a fire mountain. Hence we meet the dark goddess Pele who, like Hel, keeps the souls of the dead alive in a regenerative fire until they are ready to be reborn. Instead of terrorizing and leading to eternal torture, like the Christian notion of hell, Pele's cauldron or volcanic caldera led to the promise of eternal life. With this in mind, the next time someone tells you to "Go to hell," remember they are in fact giving you a wonderful blessing, and you may reply: "Thank you, I will! May you be so lucky!"

One other interesting fact associated with the goddess Hel is that in many traditions we learn of the lords of death who, wearing a helmet, or mask of invisibility, are able to pass through the underworld undetected by the Dark Goddess, thereby escaping permanent death and entering the paradise of rebirth. Having achieved this feat, they are often called the gems within the womb—similar to the Buddhist jewel within the lotus, which symbolizes the brilliant light of the Divine shining out from between the petals of the transformed personality.

With our understanding of the activation of the light-producing pineal gland, it is clear that this gland is the gem or jewel that causes the development of the cloak of invisibility or the Ka. With this helmet in place, the wearer is able to move with ease between different dimensions.

Artemis, the Huntress

As the Virgin, the image of Artemis is one of a free and independent spirit running through the woods at night with her hunting dogs, needing only her intuition to guide her. As the Crone, she embodies the power of life over death, for now, as the huntress, she is ready to kill the very creatures she has brought into existence. Yet her power is not malevolent but comes from a place of fate that has no connection to emotions, desires, or whims.

To Artemis, there is a time for everything: If it is time to die, we die. It is because of this clear, unattached focus that she is assigned to be the patron of midwifery: Birth is seen as a point of transition at which time life hangs in the balance, and only the Crone can decide the baby's destiny.

One of the best-known tales of this Greek goddess embraces this philosophical approach to life, although the tale has been corrupted by those who prefer to see Artemis with human emotions rather than with the detached face of fatalism.

> A young hunter comes upon the goddess bathing naked. Apparently offended by his effrontery and humiliated by her vulnerability, Artemis immediately turns him into a stag and encourages his own dogs to tear him apart until he is killed.

The myth actually records an ancient Minoan sacred ritual: It describes the fate of the stag-king god whose reign lasts for only six months before he is "torn apart" and replaced at the height of the summer solstice.

Artemis's association with fate is reinforced by the fact that she is identified with the Great She-Bear, the constellation Ursa Major, which contains the well-known Big Dipper. Her movement through the sky highlights her position as the protector of the axis mundi, or the pole of the world, demarcated in the heavens by the North Star or Pole Star. For ancient peoples, the beginning of each new season was marked by the position of the tail of the Great She-Bear, Artemis acting as the eternal clock.

The Dark Goddess is calling "time" for all of us now. She is not interested in our bargaining tactics, excuses, or entreaties, especially when these are linked to the desire to keep our personal world unaffected by the clock that is ticking and the change that is already upon us. It is time to move on . . . whether we are ready to or not.

Mah or Mut: The Vulture Goddess

For several years during meditation, I had the distinct impression of having a beak, a crown or tuft of hair, and beautiful sleek feathers. I had even "felt" myself riding the thermals high up in the clear sky, my wings outstretched, looking down at tiny moving dots on the ground. Then, about three years ago, the impressions were so strong that I decided to ask the help of Makua, a wise and wonderful Hawaiian kahuna and one of my three mentors.

He drove me to the lava tubes created by Pele's powerful volcanic eruptions and suggested that I take the winding path down to the bottom of the caldera and allow the natural environment to work its magic on me. As I descended, I opened myself to the consciousness of the bird and felt my body transform immediately until I possessed a powerful beak, keen piercing eyes, and sharp talons. Believing that I'd shapeshifted into an eagle and rather proud of the fact, I asked within: "Who are you?"

The reply took me by surprise: "I am a vulture."

"Oh no," I said, sighing, as all my prejudices rose to the surface. "How can I tell this holy man waiting for me at the top of the path that I am a vulture?" Fortunately, within seconds of that thought, my curiosity moved me to ask, "But you hover in the sky, waiting for animals to die?"

As soon as this thought left my mind, I felt the shaking of the wings of the vulture within me as they demanded respect. "Yes, you're right. I eat the dead, but not the living. I clean the bones of every animal, freeing each from its earthly attachments so that its spirit can return to the source and be reborn. Only those who fear death or have forgotten the regenerative cycles see us as ugly pests. Others welcome us with open arms."

I was aware that I hadn't always looked favorably upon this beautiful creature.

"Our eyesight is keen," the vulture continued, "and we are able to see over a much greater distance than almost any other bird. This allows us not only to see objects on the ground, but also to observe the beauty and perfect harmony of our multidimensional existence, which is so close that you could touch it and yet, for a limited human mind, might as well be in another universe."

Humbled by the wisdom of this mighty bird, my new teacher, I returned to where Makua was waiting and proceeded to tell him what had occurred. "Of course," he said as I finished. "The vulture is the condor, a bird of spirit who lives and moves between the dimensions much as Hermes moved between the worlds delivering the messages of the gods. The vulture can live in both worlds, those of the living and the dead, and is especially adapted to clearing the way so that others can follow, even if that means cleaning the meat off the bones of the dead."

Since then, I have learned that the vulture is in fact one of the oldest totem animals of the Crone and is known in Egypt as the angel of death. Several cultures, including that of the Tibetans and the ancient Iranians, would not bury their dead but rather laid them in an open-topped "tower of silence" so that the vultures could perform the last rites. Part of the reason for this was that the hard earth and desert were unsuitable for an earth-based burial, but the Iranians also built these towers to honor the moon goddess Mah, believing that the vultures would carry the dead to the heavenly realms. Even when burial was introduced by the Persians, the body was still broken apart by the vultures before it was interred.

To the Egyptians, the vulture-headed goddess was seen as the origin of all things, embodied in figures such as Mut and Isis, the latter often appearing as a vulture holding the ankh, the cross of life, in one of her talons. It is as the vulture that she tears the flesh of her dead consort, Osiris, just as Kali devours her dead husband, Shiva, and then reincarnates him in her womb (reassembles the pieces) before giving birth to him as Horus, the son and heir. Thus we see that Isis's womb provides both the power to nurture and give birth and the power to destroy, with destruction described as the sarcophagus or flesh-eating coffin.

The Valkyries: Crows, Hawks, Mares, and Swans

We meet other corpse-eating goddesses within the tales of the Valkyries, who, as attendants of the Norse god Odin, take the form of carrion-eating birds such as crows and ravens. Indeed, the word *crone* is thought to have come from *coronis,* meaning "a crow." Alchemically, the appearance of these birds in our life represents the first signs that the Dark Goddess is calling.

The Valkyries also shapeshift into hawks, swans, and mares, with the word *nightmare* arising from the belief that this frightening dream occurred after a visitation to the underworld riding on the back of the Dark Goddess. All three animals are associated with an ease of movement from one world to the next. Thus the shamans of old wore swan feathers to assist their journey between the dimensions. It is not surprising to learn that a figure of an old woman riding on the back of a swan is seen in the contours of the landscape of the Isle of Avalon, which is known as being a portal into the underworld.[3]

In alchemy, the swan represents the end of the putrefaction stage of the process, the time when we can see a milky fluid reminding us of the white light of resurrection.[4] There are many stories pertaining to the swan, with the most famous, perhaps, being that of the ballet *Swan Lake.* Written by Tchaikovsky, it tells of the powerful attraction between the heir to the throne, Prince Siegfried, and Princess Odette, who has been turned into a swan by day and woman by night, by the evil sorcerer, Von Rothbart. Even though this is a relatively modern tale and has various endings that include the romantic and the tragic, it signifies the spell that the Dark Goddess, the swan, holds over the young king, who knows he is ultimately destined to succumb to her clutches.

It is interesting to note that in this particular version of the story, the princess's alternating persona is contrary to the archetypal interpretation of the myth, which is that we are human by day and swans by night, free to fly within our dreams.

In other traditional tales, we hear how the wings of the swan are seized and hidden, trapping the creature in this physical world. Some scholars relate this story to the entrapment of a woman during her childbearing years, unable to fly until her wings are returned at menopause.

Finally, there are myths telling of the wicked witch, the Crone, who turns men into swans, signifying their descent into the underworld. Eventually, they are returned to human form by a maiden, but only after the maiden has gone through various trials to achieve their release. This story is symbolic of Inanna's descent and reminds us that it is only through the continual interplay between our Crone and Virgin (death and rebirth)—with one giving way to the other—that our masculine aspect will eventually be reborn into eternal life

All of these stories touch on the true esoteric meaning of attaining our "wings," which, as we have seen from the description of the caduceus, describes the drawing up of energy into the third eye chakra, or ajna via—our magician's wand. The two lobes of this chakra (mirroring the two lobes of the pituitary gland) fill with energy until the "wings" are powerful enough to transport our essence to the pineal gland.

Kali: The Destroyer

Even though Kali is the Hindu Triple Goddess, she is best known for her dark embodiment: She sits astride the body of her dead consort, Shiva, and eats his entrails while her yoni, or vulva, devours his lingam, or penis. More than any other goddess, Kali symbolizes the archetypal image of the birth-death Mother whose womb is also a tomb that gives life and death to her children. This poignantly reminds us that at the same moment that a woman gives birth, she is also consigning her child to the inevitability of death.

To Western eyes, Kali is often seen as the she-demon, and yet in truth, she represents the Great Mother or pure being of consciousness and bliss out of which all emerge and all will eventually return. Any projection upon her is simply a reflection of what we fear most within ourselves. Even the powerful Vishnu was said to have admitted that he was just a mere construct of her maternal substance, with her timeless ocean of blood being the source of all creation.

Mary Magdalene

Throughout Europe it is not uncommon to find cathedrals dedicated to the Black Madonna, the most famous being those in Chartres, France; Czestochowa, Poland; and Montserrat, Spain. There is some question as to who the Black Madonna actually represents; her statutes often depict her sitting pregnant or with a child on her lap. The people of France strongly believed that this is Mary Magdalene, wife of Jesus and mother of his child.

Yet if this is the Magdalene, she must be acknowledged as far more than the physical companion to Jesus. She was a high priestess, trained in the temples of Isis in the ways of the Crone, where death and rebirth were fully recognized to be the pathway to enlightenment and the elixir of life.[5] Through her "marriage" to Jesus, she offered him her womb, or cauldron, for his journey into the underworld, where he met his own demons and tempters. Here, he allowed the powers of putrefaction and fermentation, which she offered as the Dark Goddess, to work upon his psyche, tearing down any barriers that prevented him from knowing himself in his completeness. Eventually coming face to face with his demons, he found within his heart a place of acceptance and immediately became enriched by the light of consciousness, which each aspect held at its core.

Through recurrent journeys into the underworld, his light body, or Ka, became stronger until he was able to undergo physical death and, within three days, resurrect fully within his light body—the expression of immortality.

That Mary Magdalene was energetically powerful enough to be able to contain Jesus's energy as he went through this process reveals her own level of enlightenment and the fact that, prior to his descent, she too would have had to descend to meet and integrate her own demons. In essence, she is seen to represent Sophia, the wisdom aspect of the Great Mother often symbolized as both a serpent and a dove. Described as God's soul, the source of his power, and the spirit of light (Holy Ghost), Sophia is similar to the kabbalistic Shekina and the Hindu Shakti. All have received "bad press" at the hands of those who would prefer to diminish their importance, and yet without the respect and honor they deserve, there can be no sacred marriage between soul and spirit and no chance to achieve the eternal life that is being offered to us all at this time.

THE SEXUAL FIRES OF PURIFICATION

Many ancient cultures had priestesses within their midst who practiced sacred sexual rituals in order to bring about purification and transformation, both for themselves and their partners. Such maidens, whose symbol was the serpent, worked under the guidance of the great Crones such as Isis, Ishtar, and Kali, and they were revered for their mastery of these sexual fires. Through their training, they came to honor and respect the powerful serpentine energy, which runs along the spine, linking the chakras and activated by any unifying event, including sexual intercourse. Eventually, through the correct use of the breath, refinement of the energy expressed by each chakra, and a healthy relationship between their physical and spiritual worlds, they became a vessel of fire for the purification of any soul who came into their presence.

Radiating a rainbowlike aura, they provided an interdimensional staircase, allowing all those in their vicinity to raise their frequency of consciousness and thus align to their own sacred truth or blueprint. On other occasions, the maidens engaged in sexual intercourse, burning away any attachments their partners might have to a false self and allowing these partners to climb the maidens' serpentine ladder until they came face to face with their own eternal nature.

Over time, however, the patriarchy found it demeaning to men to require the services of women to connect to their own divine nature. Therefore, they cast out the sacred priestesses to the edges of society. Here, they have been waiting until now. Many men and women are awakening to the sacred nature of their sexual energy and remembering that there is nothing more loving and healing than bathing in the fires of sexual pleasure.

SYMBOLS ACCOMPANYING THE CRONE

The Cauldron

Like Sekhmet, Kali is associated with a blood lust and is often seen carrying a cauldron or pot of blood in her hands. This fiery object is believed to represent the womb of the cosmic Mother and is associated with many goddesses, including the Welsh goddess Branwen, the Celtic Cerridwen, the Irish Morgan Le Fay, the Greek Demeter, and the Babylonian fate goddess Sitis, mother of the stars.

As portrayed in Shakespeare's *Macbeth,* the cauldron is usually in the care of three witches or three wyrd (weird) sisters, symbolizing the Triple Goddess. At other times, three cauldrons are depicted. Whether one or three, the cauldron's purpose is clear: it is the container of the wise blood, the mead of regeneration, or the ambrosia of eternal life—in other words, the womb.

Throughout history we read of the daring adventures of gods who manage to steal this elixir of life from the Crone, giving them the powers of youthfulness, healing, wisdom, shapeshifting, and transformation. The Norse god Odin enters a womb-shaped cave as a phallic serpent and mesmerizes the Dark Goddess by making love to her. The god Indra is able to drink the ambrosia by allowing himself to be swallowed by the great serpent Kundalini, representing the female power of transformation, before flying away as a bird. While these tales idealize the antics of the gods, there is little doubt where the true power resides!

The Holy Grail and the Wound That Will Not Heal

According to Celtic traditions, the cauldron of regeneration is considered synonymous with the Holy Grail, the fountain of eternal abundance. In stories such as *Conte del Graal,* written by Chrétien de Troyes (ca. 1160–1180 CE), and *Parzifal* (*Perlesvaus*), written by Wolfram von Eschenbach (1220 CE), we meet the Fisher King or the Grail King and learn of his wounding by an impetuous young knight who wields a holy spear or lance, which is kept with the Grail.

The wound is in the thigh (implying impotence), which causes the Fisher King's land to be barren, and although the Fisher King continues to be fed from the cauldron, he can neither be healed nor die from his wound. Held between the worlds, his release from the fires of his eternal hell is completely dependent on the trials and adventures of his subpersonalities, the Knights of the Round Table. Ultimately, it is only Galahad, the lover, whose heart is pure enough (without judgment or conditions) to be allowed to drink from the Grail. Galahad then takes the same lance that wounded his grandfather, the Fisher King, and heals the wound. Having fulfilled his destiny, he dies.

Metaphorically, it is Galahad's willingness to own and love all parts of himself that gives him the purity and the power to wield the lance as magician, eventually bringing healing to his grandfather. And what is this lance? Often described as bleeding from one end, it is also known as the sword or spear of destiny and symbolizes the serpentine wand, which is the property of the magician that honors it. It has the ability to manifest form out of spirit and distill spirit out of form.

In the Christian legend, this lance is said to belong to Longinus, the blind centurion, who is said to have thrust the spear into Christ's side at the time of the crucifixion. The story tells how some of Jesus's blood fell upon the soldier's eyes and he was healed. Celtic teachings speak of the same spear but warn that only those of exceptional spiritual purity should attempt to wield its power. If used by lesser beings, they will receive the same wound as that which afflicted the Fisher King.

Throughout history, this spear of destiny has been linked to the belief that whoever possesses it rules the world. It is said to have been behind the successes of many great generals and leaders, including those of Constantine and Napoleon, before being placed in the Hofburg Museum of Vienna, among the Hapsburg collection, where it caught the eye and desire of a young man called Adolf Hitler, who managed to obtain it. After his death, the sword of destiny was apparently seized by U.S. soldiers and was returned to the Hofburg Museum after the war.

Esoterically, the bleeding lance represents the pillar of Jachin, which in turn symbolizes spiritual establishment, or the feminine power related to intuition. In the same Grail stories, there is mention of a sword that possesses unique or extraordinary qualities that can empower the hero with superhuman abilities. Often identified with the sword of David, this artifact is believed to represent the pillar of Boaz, symbolizing kingly strength, governance, and law.

The Grail stories were written in Europe at a time when there was a resurgence of the feminine. One thematic interpretation of them is that they warn that physical strength and reason on their own are not enough to make the hero master of his own destiny. Only when compassion and intuition are included in the hero's repertoire will the magician emerge and will our hero become such a master.

Interestingly, within one hundred years of the writing of these Grail stories, the witch hunts began in Europe. For the next five hundred years, the feminine was plunged into a darkness from which she is only now emerging into the light.

The Apple

One final symbol deeply connected to the Crone is the apple. I have many fond memories of friends and family attempting to bite into an apple, whether it was hanging from a string or floating in a tub of water during Halloween parties. At such times, we might not appreciate the strong association between these practices and the journey into the underworld in search of the elixir of life.

Although never mentioned by name in the Bible as the fruit that caused man's ejection from the Garden of Eden, it is assumed by many that the apple was the culprit. This conclusion is reinforced when we start to appreciate the deeper significance behind this Bible story. The apple is in fact an ancient symbol of immortality, appearing in mythology all around the world, including in one of the most famous of all fairy tales, *Snow White and the Seven Dwarves:*

> The wicked hag, the maiden's own dark sister, persuades Snow White to eat the red side of an apple (shed her blood) in the belief that with one bite, Snow White's dreams will come true and she will know pure love. As Snow White bites into the apple, she falls immediately into a deathlike sleep and remains there until her loving prince arrives and, with one kiss, restores her to life.

While we may assume that the moral of the story is "Don't accept strange fruit from ugly women," this simple tale tells us that it is only through our descent into the realm of death that the truth contained within our dreams and illusions will be transformed into the kiss of true love, or amrita, the elixir of life. The dwarves, with their distinct personalities, represent the seven chakras or seven gates through which Snow White–Inanna must pass on her descent into the underworld.

Thus the apple, the symbol of immortality, is found within many ancient traditions, including that of the Celts, who named their sacred site of transformation Avalon, meaning "isle of apples." To the Scandinavians, the apple was essential to ensure that humans earned resurrection; they placed the fruit in graves and inserted an apple into the mouth of a sacrificed boar during Yuletide, or the winter solstice, to ease the birth of the new sun. For the Greeks, it was believed that the goddess Hera kept a magic apple garden in the west, where the Tree of Life was guarded by her sacred serpent.

One of the reasons the apple is treated with such reverence is that if it is cut transversely, the core with its seeds forms a perfect pentagram, a magical symbol of transformation that lifts us above this physical world and into that of spirit. In other words, inside the flesh of the fruit lies the potential to be transformed and to enter the eternal world of the Great Mother, where everything is held in a dynamic state of potentiality. Yet to reach her, we must be willing to eat and be nurtured by the flesh of our own creations or experiences until only the core remains—a task that only the Crone can facilitate.

Once the apple is afforded its rightful status, the experiences of Adam and Eve within the Garden of Eden must be reexamined. It is hard for some to understand why a loving God would be happy to allow humans into his garden but not permit them to eat the fruit that would "open their eyes so that they too could become like gods" (Genesis 3:5).

As Eve listens intuitively to her serpentine, inner wisdom and eats the fruit from the Tree of the Knowledge of Good and Evil, she absorbs the understanding that it is through the cycles of death and rebirth that she will eventually know immortality. When the words *good* and *evil* are translated into the world of duality, we start to truly appreciate the serpent's message: it is only through the acceptance and mastery of the dual nature of our existence that we will come to understand the Trinity and know eternal life (the Tree of Life), which supports and is supported by these twin energies.

Although Adam and Eve were thrown out of the Garden, we are left with a key to remind us how to return: "He placed at the east of the Garden of Eden cherubim and a flaming sword which turned every way, to keep the way of the tree of life" (Genesis 3:24). One of the many possible interpretations of this is that the path to rebirth is via the east, as symbolized by the placement of the pillars of Jachin and Boaz. We must first grasp these heads of the twin powers of physical and spiritual strength. Then, through mastery of the elements (the four-winged cherubim)—air (eagle), fire (lion), water (serpent), and earth (bull)—we will manifest spirit into matter and become king. Taking this energy, we will then descend into the realms of fire, cutting away the old flesh until there remains only the truth (the flaming sword). Feeding our hearts with this energy, we will become clothed in our Ka, or light body, and thus gain access to the tree of eternal life and become as gods.

The Earthchild

There is one last esoteric factor that requires clarification before we explore the psychological aspects of the descent. Having worked with the chakras for more than twenty years, I have become increasingly concerned by the trend to number these energy centers in ascending order starting with the base, or Muladhara, chakra. Not only does this convention dilute the rich symbolism behind a name, but it also limits the opportunity to explore the presence of any of the centers located beneath the base chakra. Does it not seem strange that no attention is paid to the energy tracking along the legs and into the feet?

It is now known that there are twelve major chakras, one of which is called the earthchild, or root chakra, located about eighteen inches beneath the feet. This center, known to judges, priests, and some Scottish men—all of whom wear some type of robe or skirt—connects us to the powerful energy that runs within the earth and is seen as the Mother's blood flow. Similar to volcanic lava, this hot energy is accessible to all of us in order to bring about the transformation we hope for, not only for ourselves, but also for the world. Like Scottish men, we might become aware that when we wear a skirt, we can embody this energy with greater ease than when we wear trousers. In my case, my conscious awareness of connecting to this energy through my feet has increased, reminding me that I am resonating with the pulse of the Great Earth Mother.

THE ONLY WAY TO GO

Having gathered information from all these mythological tales and symbols, it is helpful to outline the next phases of the journey and the players involved. Embodied by different archetypes, they are repeated in cultures time and again so that we may refine the distilled essence of our soul. These include:

- **The lover:** The lover descends into the underworld to meet the Dark Goddess, whose task it is to eat away the meat of our stories until only the "bones" of experience survive. This occurs under the watchful eye of the Crone.
- **The sage:** The "bones" of our true nature are revealed with the emergence of the sage.
- **The self:** Each aspect of the self is offered to the perpetual fire of the heart for acceptance and integration, leading to the transformation of the lower frequency of form into the higher frequency of essence. This is watched over by the Virgin.
- **The magician:** This essence is fed to the ajna, or third eye, to stimulate the production of the elixir of life, the wings to fly. This is the phase of the magician.

If we return to the Crone, we see that her contributions work to strip us, like Inanna, of any regalia that supports only an outer identity, until we know the nature of the true self that lives within. The Crone also purifies and ferments us in her fiery cauldron, removing the "stuffing" of our stories until the truth becomes clear.

After this is achieved, we can then ask the following questions (on behalf of the sage):

- Why did my soul create this scenario?
- What part of my energetic blueprint did I meet that requires integration?
- What did I learn about myself during this event?

This process continually returns us to the fire until there remains only the golden nugget of truth. This process calls on us to take responsibility for those parts of ourselves that, once created, we have attempted to bury in the dark recesses of our mind (deep unconscious) due to their association with shame and fear.

For many, such a process is daunting, and yet it is important to appreciate that the prime reason for the Crone to be so demanding is that she refuses to let us be less than we have the potential to become. Much as a parent loves his or her child, her actions may appear crude and sometimes dispassionate, and yet her heart, brimming with love, fully embraces all aspects of our persona, even those parts we have so much difficulty accepting.

Now under her guidance, it is time finally to descend into the basement or cellar and break open the seals on all those storage chests and old boxes that have been collecting dust, not only during this life but for many lifetimes. Most have been moved from house to house (life to life), unopened and untouched, steeped in the fear of what may be unleashed should the contents ever see the light of day.

Yet the burden of carrying them any further is now weighing heavily, and there is no doubt that without them, our "house" would certainly feel lighter, for we know that as we unearth the true treasures contained within, our Ka, or light body, will become brighter.

ARCHETYPES AND COMPLEXES

What are these parts of the self that we are remembering? Most are what Jung called *archetypal complexes.* The word *archetype* means "first mold" or "the organizing force to create," and the word *complex* expresses how, over time, beliefs, emotions, and stories have become associated with that archetype. An analogy is that an object left at the bottom of the sea slowly becomes encrusted with corals, seaweed, minerals, and mollusks until it takes on a whole new shape, although the original item still exists within. Thus, regarding the archetype of the Mother: Woven around her are images of nurturing and compassion as well as those that portray her as being overprotective and demanding.

Every pure archetype has a variety of faces, all of which need to be expressed to be able to fully embody that particular frequency of consciousness, just as the harmonics of a tone enrich the depth of the sound. Thus we may see that the personification of the sexual woman may be expressed as a high priestess, a sacred dancer, a highly fertile mother, a seductress, a prostitute, and the victim of rape. In a similar manner, the archetype of the king can be identified with the emperor, the generous overlord, the kindly father, the omnipotent leader, the tyrant, and the one fearful of betrayal. All may be described in terms of "past lives," although it would be more accurate to say that are these "lives" are just different facets of the same holographic image.

Unfortunately, due to societal and religious beliefs associated with what is "right" and "wrong," many natural expressions of an archetype have become separated or disassociated from the core of our being: our heart. The greatest emotions relating to such abandonment are shame and fear: An individual will do everything in his or her power, often over many lifetimes, to hide the secret of his or her own essential being.

Yet however well hidden, every archetype emits a powerful frequency of energy that organizes anything in its presence into a particular pattern unique unto itself. This is the same power of sound that creates the shapes seen in the work of Hans Jenny[6] and Masaru Emoto.[7] In the case of archetypal complexes, which exist commonly outside the control of conscious awareness, their vibration attracts situations into our lives that, at first glance, seem to be unrelated to our personality.

For example, if we are disassociated from our anger, we will tend to attract angry people into our life until finally we understand that the problem is not outside but within. Like attracts like. If we take this further, we may discover that not only do we have fear around anger due to the presence of anger within our family, but also that we actually have fear and shame around our own inner tyrant—and we see his reflection in a member of the family whom we secretly despise.

Now the petal of our flower has been unfolded and the nugget of potential gold has been revealed. By taking the tyrant into our heart and accepting it as part of the self, integration occurs and the identified subpersonality is transformed into the golden essence of consciousness. It is the role of the Crone to help us reveal those aspects of ourselves held in darkness, for in truth they are all parts of the Great Mother. It is her love that can embody our deepest and rawest emotions, our most terrifying faces, and our most shameful experiences. She has no interest in judging, fixing, rescuing, or forgiving, but asks only that we reveal our authentic self and demands that we remove any mask or veil that hides the beauty of our truth.

The story of Inanna tells us that when Inanna fails to reappear, her faithful servant (intuition) seeks the help of the gods. Only Enki, the wise one, offers a solution:[8]

> From beneath his fingernails, he takes some dirt and fashions two little androgynous creatures, Kurgarra and Galatur. Because they are as small as flies, he believes they will not be seen by Ereshkigal's guards. He tells these creatures that when they enter her chamber, they will see the Dark Goddess moaning and about to give birth. He advises them to mirror her moans so that when she says "Oh! Oh! My inside," they should say, "Oh! Oh! Your inside," and when she says "Oh! Oh! My outside," they should say "Oh! Oh! Your outside."
>
> The two little creatures do as they are bid: When they are inside Ereshkigal's birth chamber, they mirror her cries. She is so amazed that they are not frightened away by how she looks that she offers them a gift. They ask for Inanna's corpse that hangs on the hook, and once the body is in their care, they pour the juice of life upon it and Inanna returns to life.

These tiny, asexual creatures fashioned from the basic component of life are apparently devoid of an agenda or personal desire and thus are able to be completely present to the Dark Queen's distress, offering her unconditional compassion. Having spent most of her life hiding in the caverns of the underworld and fearful that her rawest emotions will cause further alienation, she is amazed by their empathy and bestows upon them the gift of Inanna's rebirth.

It is this level of love that we are asked to shower upon all our archetypes that have been shrouded in shame or fear. This happens only when we, like the small creatures, fully embody their energy, feeling the range of emotions associated with such characters without the need to forgive or make sense of the situation.

This is the most challenging part of the descent, for it is far easier to understand in our heads rather than know in our hearts. On many occasions we would rather just forgive and forget, especially when deep, uncontrollable emotions arise or when we see ourselves suddenly in another person and prefer to turn away. Yet this is not a time to hide from ourselves. Instead, we must embrace our deepest shadows, a concept expressed powerfully in this poem by Thich Nhat Hanh entitled "Please Call Me by My True Names":[9]

I am the twelve year old girl,
refugee on a small boat,
who throws herself into the ocean
after being raped by a sea pirate.
And I am the pirate,
my heart not yet capable of seeing and loving.
. . . Please call me by my true names,
so I can wake up
and the door of my heart
could be left open,
the door of compassion.

It is only when we remember that we are all one and feel the heart that beats in every person we meet that will we truly know the full meaning of forgiveness. In reality:

I am just another one of you.

10

THE TRUTH SHALL SET THEM FREE

As each archetype appears out of the darkness, we may experience prejudice, fear, and shame (as I did with my vulture), causing us to seek separation from this aspect and to experience a desire to offer it forgiveness in order to rid ourselves of its presence. Yet the only way through is down, delving deeper until we know the archetype in its purest form and we can embody its energy without fear. This is not an easy process, and yet without our ability to say "I accept you as part of myself," we deny ourselves the opportunity to enter the heart of the Great Mother and experience the abundance that waits beyond.

It is the sage or wise man within all of us, symbolized by the sign of Sagittarius, who knows the truth. With detached compassion, he meets each aspect or subpersonality as it emerges from the fire, seeking an authenticity rather than judging its worth. In alchemy, the completion of the process of fermentation is symbolized by the beautiful tail of the peacock. This reveals the many "eyes" of our subpersonalities which, liberated from the caverns of despair and rejection, we are proud to display for all to see.

♐ SAGITTARIUS

Quality: mutable; focused awareness or the truth

Alchemy: the beginning of distillation

Polarity: masculine; sage

Last or Third Quarter Moon Phase: to revise and reevaluate

Throughout my life, I have met many of my subpersonalities wrapped in what we call past-life stories. On most occasions, I did not go in search of them; rather, they came and found me, bringing enormous opportunities for growth and spiritual empowerment. Thus, over time, I have met myself within many forms:

- A happy Irish mother with eight children
- A Roman soldier who hangs himself in conflict over his belief in Jesus's teachings and the orders given him by his superiors
- A black servant given his freedom
- A pioneering farmer exhausted and despairing of the continual rains that wash away the crops
- A brave Native American warrior devoted to his family and to the land
- A child who falls into a pond and drowns
- A tyrant and soldier who treats everyone with contempt and who dies alone and in pain on the battlefield
- A female black magician who uses her skills to possess what is not hers to possess

As these character have appeared to me, each has brought an emotional quality that energized every cell of my body until I felt that I *was* that subpersonality. Such embodiment brought with it authenticity—the knowledge that all the "meat" had been removed from the story and I was experiencing the core consciousness.

This was not always an easy process, especially in the case of one of my more recent connections, when I tapped into a cruel, cold, and calculating energy. It was very clear from the start that this subpersonality inhabited the lower three chakras of my body, preventing me from feeling fully secure, nurtured, and confident. Initially, I offered it love, partially in the hopes that it would leave me alone, but it only sneered at me. "I am unaffected by your humanoid emotions," it said.

Many weeks passed, and I was growing desperate, for, like Lilith in Inanna's tree, this energy seemed to have made my body its home. Calling out for help during a seminar, I heard the advice: "Draw into the heart those parts of the self that cause distress, and say to them: I accept you as part of myself." On following these instructions, the energy changed immediately and the reptilian energy—for that is what it was—became absorbed and transformed within my heart into the pure essence of consciousness, judged as neither good nor bad, but rather as an essential part of my soul.

It is the role of the sage within us all to question our deepest intentions and beliefs until only the truth remains. Such a process is reflected in the story of Inanna's descent through the seven levels. For the Sumerians, the number 7 was significant, for it was seen as the number of completeness. It reminds us that the Virgin, our spiritual blueprint, reaches her radiant wholeness only through her courage to meet herself in all the darkest corners of the psyche.

Hence, we are asked:

- What is your true authority?

  This relates to the first gate, the crown chakra. Do you need a physical crown to claim sovereignty of your being? It is easy to recognize a lack of self-esteem when there is a need to exhibit superiority, arrogance, pride, or defensiveness. It is only without the crown of external success or pride that we come to know true authority.
- What do you not want to see?

  This relates to the second gate, the third eye. Lapis lazuli opens the third eye, bringing to its wearer an enhanced ability to journey into other realms, increased psychic awareness, and expanded spiritual growth. It also encourages us to turn our attention from the outer world to the inner and face ourselves squarely in the mirror.
- Can you let go of control and trust the outcome?

This relates to the third gate and the throat chakra. At this gate we surrender the need to dictate the outcome through bargaining, analyzing, and excuses and learn to flow with the journey. This is probably one of the hardest gates.

- What is left when there is only love?

  This relates to the fourth gate and the heart chakra. When we release conditions, expectations, demands, and desires, what is love?

- Who are you?

  This relates to the fifth gate and the solar plexus. At this gate, we are guided to let go of attachments that support an external identity and reconnect to the confidence that emerges from the core of our being.

- Why do you hide?

  This relates to the sixth gate and sacral chakra. This gates connects us to our relationships and asks where we are hiding due to shame and humiliation and where have we failed to respect ourselves. Ereshkigal demands authenticity; hence, there is no place for shame or secrets. When we place the secretive parts of our being in the Great Mother's love, the reason to hide dissolves.

- Do you belong to yourself?

  This relates to the seventh gate and base chakra. This final gate strips us of our last defenses until we are naked. It asks if we can be secure in just being ourselves.

As Inanna stands before Ereshkigal, the dark sister sees that Inanna is still wrapped in the skin of mortal "life," and with one stroke, she removes the outer covering so that the spirit can be free and she hangs the dying flesh on a peg to rot.

THE PERPETUAL FIRE

Capricorn is one of the most complex signs of the zodiac, despite its exoteric links to rules, responsibility, and loyalty. Esoterically, its glyph represents a synthesis between an animal and a fish, symbolizing the transformation that occurs at this level between our physical and spiritual natures. Alchemically, this stage is called *distillation* and is represented by the pelican, which is willing to feed its young with its own blood by pecking its breast. This symbolizes our willingness to feed ourselves with the fruits of our endeavors or gems of wisdom, recognizing that in so doing, we are willing to sacrifice our archetypal king for the promise of new life.

♑ CAPRICORN

Quality: cardinal; conscious awareness transformed into pure essence

Alchemy: distillation; the pelican

Polarity: feminine; the Triple Goddess

Balsamic Moon Phase: to distill and transform

In other words, while an old story remains and there is still energy and emotion attached to it, we will continue to leak energy back to that situation, continually causing us to feel spiritually undernourished. It is only when we are willing to learn from our experiences and be accountable for the part we played in the creation of that story that we can then extract the gems from each subpersonality involved and feed ourselves with the richness of its life force and blood so that the old can truly die.

For a deeper understanding of this process, we will once again call upon our trusted Virgin.

The Vestal Virgins

During the era of Roman civilization, six beautiful maidens were chosen to dedicate their lives to keeping the perpetual fire alight, acknowledged to be the mystical heart of the Empire. Named after the Roman Goddess Vesta, the goddess of the hearth, they were part of a much older order of priestesses who were known for their magic, motherhood, and for being those who chose the rulers. Indeed, the first vestal virgin was the goddess Rhea Silvia, who came from Crete and, as legend tells, was the mother of Romulus and Remus, the founders of Rome.

Vestal virgins

These young virgins received many privileges that were normally out of the reach of ordinary women. In return they were expected, on pain of death, to focus all their attention on maintaining the fire, remaining unmarried, and being unavailable to any other distraction.

Why would such a powerful patriarchal society bestow so much importance on a group of girls? What was so unique about this perpetual fire? During my research, I came across this quote by the Sufi master Hazrat Inayat Khan that helped to bring clarity to this important question:

> If love is pure [and] if the spark of love has begun to glow, then there is no need to go somewhere to gain spirituality, [for] spirituality is within. One must keep blowing the spark till it turns into a perpetual

> fire. The fire-worshippers of old did not worship a fire that went out, they worshipped a perpetual fire. Where is that perpetual fire to be found? In one's own heart.[1]

Vesta, Hestia (the Greek equivalent), and the vestal virgins were not merely physical fire keepers but were also protectors of the hearth, considered to be the heart (heart-h) of the home or the culture as well. The Romans knew that the perpetual fire was the source of their continued prosperity and abundance and that without it, creativity would become stagnant, infertility would increase, the land would become barren, and the people would be depressed. In essence, their empire would not survive.

Most people today are still drawn to the warm glow of a fire, recognizing its ability to encourage a community spirit, rekindle dreams, and add a magical quality to stories told. Yet so many homes are built without a central hearth or communal cooking area, and somehow a microwave just isn't the same! Hence we see people gravitating toward places where there are bright lights, such as malls or shopping centers, bonfires, and even around candles in an attempt to experience the heart of the perpetual fire.

The Heart and the Torus

To take our exploration of the perpetual fire to a deeper level, it is first important to study the facts that are already known about the heart, this unique organ of the body.

The heart is a complex electromagnetic system that, with every beat, produces enough energy to power a small electric bulb. Its amplitude is forty to sixty times greater than that of brain waves. The field produced by this energy radiates out some twelve to fifteen feet beyond the body, and the most powerful part of the spectrum exists within the first three feet. It is no wonder that we feel so good in the presence of someone who exudes heartfelt joy.

Research has shown that a single cell taken from a heart has an innate memory that allows it to continue to beat even when it is disconnected from the nervous system. In time, it will start to fibrillate and then slowly die. Yet if we take two cells that are both fibrillating and bring them together, they will at some point connect energetically and both will be restored to normal health. If two unhealthy cells in a laboratory respond in this way, exemplifying the healing power of a loving relationship, imagine the potential for two people whose hearts are open to each other to come together. Imagine what could happen to whole nations.

In another experiment, the fibrillating cell was brought into the presence of a healthy cell. Within minutes, the healthy cell entrained the sick cell back to health, leading both to beat in harmony. This power of the heart cell to entrain is based on the principle of congruency: each cell holds the memory of oneness. In this way, great healers radiate such a strong state of harmony and congruence that anybody in their presence naturally realigns to the healers' state of unity and health.

It has now been shown that the electromagnetic energy of the heart takes the shape of a *torus* (a toroid shape).[2] Geometrically, a torus is formed by rotating many circles around a tangential line, the center of which touches all the rotated circles exactly. There are three main torus shapes, the most common being similar to a ring doughnut: a tube torus.

Given the growing general interest in sacred geometry, this sacred design is currently receiving considerable attention. The Galactic Center to which we are now aligned is in fact a black hole shaped like a torus. Just as our heart is a torus, so, apparently, is the heart of the Great Mother!

Tube torus

As our research deepens into the nature of the torus, we find that it has one unique function: it is a transformer that is imprinted with creation's blueprint and as such, allows for the formation of matter from spirit and the dissolution of matter back to spirit.

In other words, it is the heart—of a human or of the Great Mother, as found at the Galactic Center—that turns spirit into matter and then converts matter back into the essential nature of spirit. It is our heart, acting as a transformer, that makes immortality possible.

When we transpose the transformational quality of the toroidal heart to our relationships, it becomes clear that it is only through the heart that we can meet people "where they are," shifting our frequency until we find one that is mutually acceptable. Attempting to achieve the same result through the head often leads to patronization, arrogance, and self-delusion.

At the same time, it is through the power of love that we are transported outside the confines of time and space. Who has not been in love and found that everything in our environment appears beautiful, blissful, and loving? Who has not found that love makes time stand still? Love can make things happen (turn spirit into matter) or help us release our hold on something and enjoy the ride (turn matter into spirit).

This sacred design has another important feature: Once energy is generated and set in motion, the torus is self-perpetuating; it maintains its own momentum. In other words, the fire is kept alight by its own desire to love. Is this not true in our own lives? We seek the intimacy of connection, which could be connection to another person, nature, our inner being, or the Divine. Every human seeks union. It is an impulse that drives all hearts—ours and the Great Mother's. This eternal need to connect reinforces the medical literature that states that isolation, lack of community, and fear of speaking from our heart are major causative factors in heart disease.

Further, what do we know of the flow of energy within the torus that creates this self-perpetuation? Once again we hear the same message: The flow is kept in continual motion by a healthy balance between equal forces of attraction and repulsion, for these two forces work together. These are the twin pillars or the energies of the ida and pingala that sustain the central flow.

This reminds us that our heart accepts all parts of the self equally, without judgment or favor, for dark and light are equally acceptable in maintaining the balance. Energy passes through the center of the torus as a swirling vortex, shifting frequencies from one level to another. This core represents the union between the opposing forces, the sushumna, which acts as the wormhole or passageway into the vastness of the Great Mother.

It is through love of the duality—unity through diversity—that we will know our immortal and eternal self. The flow within the torus is bidirectional: it enters and leaves at the top and the bottom, reminding us that spirit and matter, heaven and earth, are equally valuable and interdependent. "As below, so above; as above, so below."

HEARTFELT COMMUNICATION

One of the most common images of a toroid is as a doughnut-shaped ring whose surface consists of seven distinct colors, all of which are in touch with each other and which appear to be able to communicate with each other despite their different frequencies.

This transmission of information is known to occur not only due to a physical connection, but also energetically, using the medium of ether or nonlocal reality. Could this be the impulse behind the ability to speak in tongues? Could it be that when we listen through our hearts, we can understand each other without the need for words? Could this be the force that inspires telepathic communication?

In James Twyman's book *Emissary of Love*,[3] he questions some of the fifth-world children (also known as the crystal children), about their highly developed psychic abilities. In unison they say that such gifts are only the by-products of the power of love. They and others remind us that when we have no need for secrets, and shame has been transmuted into love, then we too will achieve instant communication, not only with each other, but also across the universe, where time and space do not exist.

Once we understand that it is through our hearts that we can communicate anywhere in the world, we understand how we can send healing thoughts to reach a loved one thousands of miles away. At the same time, it becomes clear that it is through our individual hearts that we connect to the universal heart and know our part in the great design. In fact, wherever there is a torus, there is an instant connection to the unified pulse of the source, the Great Mother. It is now clear that there are many examples of toruses within our world, and all have the ability to communicate to each other directly:

- **The atom:** The movement of the particles within the atom are toroidal.
- **DNA:** The double helix is now known to act as a torus, transforming energy and information into and out of the DNA.
- **A chakra:** The center of the torus is a vortex connecting what appear as swirling trumpets radiating out of the front and the back of the chakra. This same shape could also be described as a chalice with two openings through which energy flows in and out. Could this be the origin of the Holy Grail and the source of perpetual or eternal life?
- **The electromagnetic field of the body:** Energy pours into and out of the crown and base chakras, and the spine is the central axis.
- **The Tree of Life:** The roots and the branches exist at opposite ends of the axis.
- **The earth:** Its Van Allen Belts consist of magnetically trapped, highly energetically charged particles aligned to the north-south axis.
- **Sacred Sites and Crop Formations:** Megalithic sites such as Avebury in the south of England were built surrounded by a moat. This tubelike construction and the spiraling shape created by the stones form a torus designed to focus and transform archetypal

energy that enters the site during specific astronomical events. This energy is then sent out through the grid system of the earth to affect people's consciousness.

- **A rainbow:** A rainbow blends seven different frequencies of light within a bow, the symbol of the Virgin. This natural torus, associated with a magical pot of gold at its end, reminds us of our perpetual and eternal connection to the source and the natural dance between the Great Mother's water and the fire of the focused mind.
- **The sun:** Its energy field is toroidal.
- **The Galactic Center:** The Galactic Center is a toroidal-shaped black hole.
- **The center of the universe:** Similar hypothetically to the Galactic Center.
- **Every human heart:** Our hearts connect, even when our minds fail to listen.

Can you appreciate the communication that is even now taking place between your heart, a cell, and the center of the galaxy, programmed as they are by the same holographic view of oneness? This is why joy is the only human emotion that produces the same waveform wherever it is measured around the world. Joy links us as a species, joy extends our horizons beyond the physical, and joy allows us to pass through the heart of the Great Mother and into her abundant ocean of possibilities.

We will never know whether the Romans understood that their fire, watched over by the vestal virgins, reflected a powerful geometric design of transformation. What *is* clear is that the rulers of this mighty Empire did believe in a link between the perpetual fire and the eternal source of long-lasting prosperity and indisputable power. At this time, such gifts are available to us all as long as we follow the messages of the torus and the heart:

Let go of judgment.
Be compassionate.

Become your authentic self.
Live in the present.

THE MAGICIAN'S WAND

Known as the water carrier, the astrological sign of Aquarius is actually an air sign, signifying the final distillation of the light of consciousness from the transformation that takes place in the heart. Here the magician has been patiently waiting while, through the ascending and descending activation of the chakras, we increase the power within each strand of his wand:

- The ida through which energy is poured to create matter. This represents the ascending pathway between base and crown, the building of the ego, mastery of the four elements, and hero's journey from puer to king.
- The pingala through which we pass to dissolve form back into spirit. This represents the descending pathway between crown and base and the journey of the lover and the sage as they pass into the underworld in order to remember and integrate.
- The sushumna, the union of the ida and the pingala and the channel for the distillation of pure spirit, the elixir of life. This represents the emergence of the magician, who can manipulate the multidimensional energies with just one shake of his wand.

The energies of the pingala, which are distilled from the heart, rise up and are collected in the ajna, or third eye. Here, they meet the energies that are formed at the time of the king's crowning, the ida. Thus, the head of the manifested serpent and the head of the spiritualized serpent face each other, creating the two wings of the ajna and forming the sign of infinity.

The sign of infinity

♒ AQUARIUS

Quality: fixed; group consciousness

Alchemy: distillation and coagulation

Polarity: masculine; the magician

Balsamic Moon Phase: to distil and transform

It is through this union that the Djed is raised, the Ba is drawn down to create the release of amrita, the Ka is illuminated, and the oneness of immortality achieved. As we stand poised between the worlds, it is time to remember our eternal nature and the fact that we were all probably here on this planet twenty-six thousand years ago, during the last great shift. Each of us is unique; each carries a specific part of the jigsaw. It is time to put the final pieces into our own personal puzzle, and together we will travel through the wormhole into the heart of the Great Mother and beyond.

What a wonderful time to be alive!

11
THE EMERALD TABLET

To complete this exploration of immortality in which so much of the process is alchemical, it is fitting to study the treatise that embraces this science of the mystics, the Emerald Tablet. Here is what the Count of St. Germain, a master alchemist born around 1560 CE, says on the subject:

> The inner meaning of alchemy is simply "all-composition," implying the relationship of all of creation with the parts which compose it. Thus alchemy, when properly understood, deals with the conscious power of controlling mutations and transmutations within Matter and energy and even within life itself. It is the science of the mystic and the forte of self-realized man who, having sought, has found himself to be one with God and is willing to play his part.[1]

St. Germain taught that self-mastery was the key to this Great Art in which the alchemist could, through his actions, determine the design of his own life's creation and hence fulfill his destiny. As mentioned earlier and is worthy of repeating here, he warned any developing alchemist to be aware that self-delusion and rationalization are two of the greatest challenges to be faced.

For thousands of years, alchemists have considered the Emerald Tablet to be their bible. It is an ancient artifact said to contain advanced spiritual technology describing the steps required to achieve personal transformation and accelerated species evolution. Encompassing all levels at once—mind, body, and spirit—the tablet's teachings have threatened many who wish to maintain their dominance over the masses, causing people to be suppressed by church and state alike.

Its message, however, is greater than the small minds of men, and through time, its thrust has become enriched, translated, and revered, carrying its readers to ever-expanding levels of consciousness. Described as being molded out of a single piece of green crystal, it has caught the attention and imagination of many scholars, writers, and scientists, including Sir Isaac Newton and Carl Jung. It is seen as the original source of Hermetic wisdom and alchemical philosophy.

The tablet's whereabouts today are a mystery although some say it will be located eventually near or under the Great Pyramid of Giza. Many of its teachings are allied closely with those found in Buddhism, Taoism, and Hinduism and are the basis of philosophies that underpin Islam, Judaism, and Christianity.[2] For seventeen hundred years it has been the inspiration for most alchemists from all walks of life, including the Freemasons and Rosicrucians, although its origins are somewhat obscure. It is usually attributed to an author known as Hermes Trismegistus—but who was he?

HERMES TRISMEGISTUS

His name means "thrice greatest," which recognizes that, as far as we know, he incarnated three times.

Thoth: The First Hermes

The first Hermes has a very mixed heritage, combining history, mythology, and mysticism. According to the nineteenth-century French scholar Artaud: "Thoth, the Greek Hermes, was the symbol of the Divine Mind, incarnated Thought, the Living Word, the Logos of Plato and the Word of the Christians."[3]

Known as the masculine equivalent of Ma'at, the central beam between the pillars of Jachin and Boaz, he has been described as an immortal Atlantean priest-king ruling an ancient colony of Egypt from 50,000 to 36,000 BCE. According to researcher Dr. M. Doreal, before Thoth departed this earth, he built the Great Pyramid of Giza and secretly stored ancient artifacts, texts, records, and instruments of Atlantis within its structure.[4] Originally, there were twelve emerald green tablets formed from a substance created through alchemical transmutation, making them imperishable and unchangeable. Engraved into their surface were characters in an ancient Atlantean language, which responded to attuned thought waves imprinting the appropriate vibration directly into the mind of the reader. The tablets were apparently fastened together with hoops of golden-copper alloy and were suspended from a rod of the same material.

Beneath the pyramid, Thoth constructed the great halls of Amenti, representing the underworld, where all souls will pass after death to be judged and where Thoth's spirit waits between incarnations.

From all accounts, it is hard to say whether the first Hermes, Thoth, was ever fully human, although it is believed that he combined the multiple tablets into one text, which is known to modern occultists as the

Thoth, the reckoner

Emerald Tablet. This Egyptian deity enjoyed popularity from 2670 to 2205 BCE and was seen as the god of magic; inventor of writing (in particular, hieroglyphics); teacher of logic and speech; founder of mathematics, science, and medicine; and, in essence, the representative of One Mind.[5]

In his other role, he was known as the revealer of the hidden, the lord of rebirth, and the great measurer or reckoner of the universe. As such, he was given the authority to judge the dead in the hall of Ma'at. Here, he would weigh an individual's heart and assess how closely he or she had followed, through words and actions, the innermost intent.

In art, Thoth is depicted as the god of the moon; his head and the beak of an ibis symbolize the crescent moon or the heart. He is also depicted as a baboon, a night animal that greets the return of the sun in the morning just as the moon welcomes the sun.

Akhenaten: The Second Hermes

The pharaoh Akhenaten, also known as Amenhotep IV, ruled from 1364 to 1347 BCE, setting up a monotheistic religion much to the displeasure of the priests who enjoyed the power brought by worship of

Akhenaten with the solar disk

many deities. It is believed that Akhenaten found the Emerald Tablet at the beginning of his reign and espoused the concept of living in truth, with this universal ideal relating to the original will of the One Mind and the physical expression of the One Thing, seen as the physical sun or solar disk whose rays radiated to all humankind.[6]

Married to the beautiful Nefertiti, he too was beautiful and some say even extraterrestrial in appearance. After seventeen years of reign, during which time many of the corrupt practices of the past were overturned, husband and wife both disappeared and their bodies were never found. They were replaced by the boy pharaoh Tutankhamen, and power reverted back quickly into the hands of the priests, although Egyptian supremacy was never the same again.

As has been discussed in an earlier chapter, if Moses and Akhenaten were the same person, then perhaps Akhenaten didn't die but instead carried his message and legacy out of Egypt and toward the Promised Land.

Balinas (also known as Apollonius of Tyrana): The Third Hermes

The next time the tablet surfaces is in 332 BCE, when it is found in a temple in Siwa by Alexander the Great after his victory in Egypt. Immediately, he translated the text into Greek and used its knowledge to enhance his power, success, and fortune. He hid the tablet for safekeeping and died during a journey to India. The tablet was found three centuries later by a youth called Balinas, who was born in 16 CE in what is now Turkey.[7]

Through appreciation and integration of the tablet's teachings, Balinas became a great mystic with wise and magical powers, especially in the field of healing. Unfortunately, the followers of Christ were jealous of his powers, and by 400 CE many of his books and temples were destroyed. His writings, however, were not completely lost, and in 650 CE *The Book of Balinas the Wise on Causes* appeared in Arabic.

ALCHEMICAL TRANSFORMATION

What follows is a modern translation of the encoded teachings from the Emerald Tablet. In essence, this portion of the Tablet summarizes the production of the philosopher's stone or elixir of life, also known as the ambrosia of the gods and said to ensure that not only spiritual transformation but also immortality could be achieved. Once the process begins, each stage is associated with one or more element, chakra, and symbol. Following the entire quoted translation, its pieces are broken down with my commentary on meaning.

> In truth, without deceit, certain, and most veritable.
>
> That which is Below corresponds to that which is Above and
> that which is Above corresponds to that which is Below
> to accomplish the miracles of the One Thing.
> And just as all things have come from this One Thing through the meditation of the One Mind,
> so do all created things originate from this One Thing, through transformation.
>
> Its Father is the Sun, its Mother the Moon.
> The Wind carries it in its belly; its nurse is the Earth.
> It is the origin of All, the consecration of the Universe;
> its inherent Strength is perfected, if it is turned into Earth.
>
> Separate the Earth from Fire, the Subtle from the Gross,
> gently and with great ingenuity.

It rises from earth to heaven and descends again to
earth, thereby combining within Itself
the powers of both the Above and the Below.

Thus will you obtain the Glory of the Whole
Universe.
All obscurity will be clear to you.
This is the greatest Force of powers because it
overcomes
every Subtle thing and penetrates every Solid thing.

In this way was the Universe created.
From this comes many wondrous applications,
because this is the Pattern.
Therefore am I called Thrice Greatest Hermes,
having all three parts of the wisdom of the Whole
Universe. Herein have I completely explained the
Operation of the Sun. [8]

In truth, without deceit, certain, and most veritable
Free of dogma, without ego, centered and most intuitive.

These few words set the scene for a profound experience and the study of them is, in itself, a lifetime's work. They state that the process of alchemical transformation can occur only when we learn to:

- Master (not suppress) our emotions and energies.
- Carry a level of self-confidence that has no need to judge others out of fear that they may expose a part of us that is hidden in the shadows due to shame or fear.
- Release our need to judge or carry biases in order to protect ourselves.
- Come from a place of discernment detached from the outcome.

- Think, act, and speak from our central core, where head and heart are aligned.
- Be guided by an intuition, driven not by fear but by love.

> That which is Below corresponds to that which is
> Above and
> that which is Above corresponds to that which is
> Below
> to accomplish the miracles of the One Thing.
> And just as all things have come from this One Thing
> through the meditation of the One Mind, so do
> all created things originate from this One Thing,
> through transformation.

This rubric reminds us that life is cyclical, with no true beginning or end, and that without such a continuum, alchemy cannot work. It also allows us to understand that both spirit and matter, the above and the below, demand equal reverence, for they exchange energy continually. Finally, it confirms that when the One Mind, our attention, is clearly focused on the One Thing, our imagination, then miracles will occur.

> Its Father is the Sun, its Mother the Moon.
> The Wind carries it in its belly; its nurse is the Earth.
> It is the origin of All, the consecration of the Universe;
> its inherent Strength is perfected, if it is turned into
> Earth.

Having set the scene, the process is now described in detail, starting with the four elements that are considered to be the Prima Materia, or First Matter, for an alchemist.

Fire: inspiration, wands, clubs; calcination
Water: feeling and emotions, cups, hearts; dissolution

Air: thinking, swords, spades; separation
Earth: sensation, pentacles, diamonds; conjunction

Its Father is the Sun . . .

This phase of calcination relating to fire represents a deepening of self-actualization and a new beginning on the hero's journey. Due to the cyclical nature of existence, each of us will experience this at a different level, with some acknowledging it as a time to separate from ancestral beliefs and dogmas as the first step in leaving "home" and others recognizing the need to burn away their illusionary beliefs about reality and realign more deeply with their spiritual intention.

In essence, the questions are the same: Who is driving the bus? Who is running the show? Who is focusing attention?

From the text, the sun represents our conscious awareness as a reflection of the One Mind and asks that we allow its heat to burn away—or calcinate—any beliefs that do not align with the soul's deepest truth. As we focus our attention on the shadowy yet influential constructs in our minds, fear, excuses, and anger often arise in an attempt to defend what we would rather remained hidden.

This can be a difficult stage in our lives, especially when we have dedicated so much energy to the outer world only to see it dry up under the sun's glare. This may be experienced as a loss of possessions, failed love, depression, and loss of self-esteem. The greater our attachment to the past, the greater the amount of anger and frustration we express toward an external authority who seems bent on ruining our life until we learn to use that same fire to ignite the passion of our soul.

Symbol: black bird and fires
Chakra: base, associated with security and stability

. . . its Mother the Moon

Now the ashes of calcination are scattered on the waters of the unconscious, irrational, and chaotic parts of our mind. This dissolution, cleansing, and baptism breaks down any artificial constructs of the psyche, freeing the energy that has been trapped by a false separation between our outer and inner lives, our personality and our soul. For those who have been well defended, this phase can be frightening, for the soul's presence demands that we let go of control, break unhealthy habits, allow feelings to flow, and accept the discomfort of not knowing.

If we do not submit willingly, then our soul will create the same result through illness or crisis; ego-based expectations and long-term plans will dissolve in an instant. Initially, there are thrown to the surface all manner of fears, wounds, and pains, which may dissuade the hero from going further. Yet, these raw energies are just part of myriad emotions and feelings now available to us as we shift our attention from the dried-up landscape to the ocean of possibilities within our own imagination.

As the emotions settle, a state of euphoria may emerge; we may find ourselves floating in our own unconscious without the fears that have entrapped us for so long. Yet it is also important at this stage not to become lost in delusion, believing that this state is the eagerly sought-for union with the Divine. The dissolution into the Great Mother, the One Thing, is not an ending but the opportunity to reclaim the power and potential readily available to every one of us.

Symbol: black bird, water, mirrors, and tears
Chakra: sacral, associated with relationships and connection.

The Wind carries it in its belly . . .

As we move into this phase, it is time to separate, reclaim, and integrate the dreams and golden insights from our soul that surfaced during dissolution but which had been buried far too long due to allegiances to the "false idols" linked to shame and fear. This invigorating fresh air encourages us to listen to our intuition and apply healthy filters, choosing beliefs that nurture our soul and releasing those that enslave an unloving alliance to familial, cultural, and societal dogmas.

Over time, we feel less burdened and increasingly self-confident, as if we have been given wings to fly. We also find that our ability to be objective and clear-sighted grows stronger, helping us to avoid the tricks and illusions that attempt to draw us back into our old patterns. Now we are ready to create a new reality based on the richness of our soul.

Symbol: black bird, air, and swords
Chakra: solar plexus, associated with self-confidence

. . . its nurse is the Earth

The earth offers the vessel for the planting, nurturing, and blooming of our dreams as, from the core of our inner essence, we commit to living truth. This is a time for conjunction, or the sacred marriage of opposites, a phrase coined by Jung. It is a time when spirit and soul meet and produce a fragile child called the lesser stone. This child is the king who we met in earlier chapters. His crowning is often falsely perceived to be the end of the journey and yet merely represents the creation of a vehicle for the deeper work still to come.

It is common for this stage to be marked by an increase in psychic awareness and to be accompanied by synchronicities as we reconnect to the larger, multidimensional picture. This stage is also one in which many choose to stop, happy with their spiritual successes and not yet ready to descend to meet their demons—for the next stage requires a willingness to engage in self-reflection, self-examination, and the eventual disintegration of the smaller self in service to the greater self.

In truth, however, any attempt to avoid the call of the Dark Goddess is not a wise move, for she will demand your attention eventually, and the call may come more suddenly and more dramatically than you would choose.

Symbol: cockerel, lovers, weddings
Chakra: personal heart, associated with acceptance of duality

> Separate the Earth from Fire, the Subtle from the
> Gross,
> gently and with great ingenuity.

This stage, one of putrefaction and fermentation through fire, is the turning point in the Great Art. Commonly, it envelops us in a dark night of the soul as we choose to go within and descend to meet the Dark Goddess and our own personal demons. At some point we step into her cauldron, the vessel used for the processes of putrefaction and fermentation, and find ourselves stewing in our own creative juices. As she stirs the pot, the meat of our stories, which contains both nutrients and waste matter, becomes cooked and begins to dissolve, revealing the bones of truth.

As more of the mental chatter, emotional manure, and illusionary insights are broken down, the fermentation process brings to the surface all of our subpersonalities, which require acceptance and integration. As each is brought out of the shadows and into the light, we experience a strength and freedom that we might have thought previously to be unattainable.

For an alchemist, the sign that putrefaction is reaching its end is the appearance of a white milky liquid on the surface of rotting material that esoterically represents the white light of resurrection, proving that consciousness has survived the "death." Symbolically, this is represented by a White Swan, gliding across still waters that run deep, reflecting the fact that, despite the dark times, the inner light is always present.

The sign that fermentation is almost complete is seen in the formation of the Peacock's Tail (an iridescent film of oil that appears on the putrefying organic material), the eyes on the feathers representing the many aspects of the Self now revealed.

Chemically, at the end of fermentation, an alchemist sees the production of a solid, yellow ferment representing the final stages of the transformation of gold from the base material. This is known as the Golden Pill. This yellow, waxy substance is the literal incarnation of thought and the first indication that we are making gold. It is formed from the union of inspiration above and imagination below (One Mind and One Thing) and is called the secret fire. It opens the portal to higher realms and multidimensional existence.

Symbol: end of putrefaction, the white swan; end of fermentation, the peacock tail
Chakras: throat chakra, the willingness to enter the cauldron

From the throat chakra we repeatedly travel down through the heart to the lower chakras where we bathe ourselves in the energetic "soup," drawing back each shadowy aspect into the transpersonal heart for acceptance and integration; the peacock tail.

> It rises from earth to heaven and descends again to
> earth, thereby combining within Itself the powers
> of both the Above and the Below.

Now it is time for each character or subpersonality to be transformed into its essential energy through the process of distillation. This takes place in the heart, the perpetual fire where the "matter" of our creations becomes spirtualized.[9] Thus, distillation asks that we feed our Spirit with the wisdom of our own creations. Psychologically, this represents the dematerialization of the psyche as the individual sacrifices their earthly nature, including their base emotions and ego-centered beliefs. This process occurs between the lower and upper worlds until all value has been extracted and embodied in the form of wisdom.

During this time it is useful to watch for:

- Repeated patterns within our lives that offer a clue to their source.
- Mirrors of ourselves in others often revealed in the form of an emotional reaction.
- A desire to defend ourselves or judge those who get too close to the truth about who we are.
- Stories we tell ourselves, which are irrationally rational.

Psychologically, we will find ourselves moving between the subtle and gross aspects of our personality until we find peace and well-being. At the same time, as we integrate the essence into our own light body, or Ka, we start to create a glowing magnetic field, enhancing the attraction of those people and events that maintain our state of harmony.

Symbol: distillation represented by the pelican, which feeds its young with its own blood—our ability to embody the consciousness of our experiences

Chakra: the heart and the ajna

> Thus will you obtain the Glory of the Whole
> Universe.
> All obscurity will be clear to you.
> This is the greatest Force of powers because it
> overcomes
> every Subtle thing and penetrates every Solid thing.

Coagulation occurs when the distilled essence rises up to meet the essential self or Ba and the second sacred marriage occurs to give meaning to the phrase: I am that I am. Once this stage is reached, we know the freedom of spirit that can move with ease between the dimensions. For the alchemist, this mobile state of consciousness is the philosopher's stone, which we know also to be the elixir of life.

Symbol: the phoenix
Chakra: the ajna and the crown chakra

> In this way was the Universe created.
> From this comes many wondrous applications,
> because this is the Pattern.

This pattern is available to everybody

> Therefore am I called Thrice Greatest Hermes,
> having all three parts of the wisdom of the Whole
> Universe. Herein have I completely explained the
> Operation of the Sun.

This explains the principle forces that are essential for the process of alchemy and enlightenment and symbolized in the Trinity. The same message is reflected in other trinities, such as the inter-relationship between the pillars of Jachin and Boaz linked by the central beam Ma'at, and the relationship of the Virgin and Crone to the centrally based energy of the Mother. It is through a healthy relationship between the outer two elements that the central beam continually agrees to feed and nurture their existence. The one gives birth to the two and through their sacred marriage of opposites, the needs of the one are eternally met.

Spirit	Child	*Soul*
Spirit	Mental	*Matter*
Sun	Moon	*Earth*
Sulfur	Mercury	*Salt*

12

THE LUNATION PHASES AND THE NODES OF THE MOON

The famous astrologer Dane Rudhyar first examined the lunation phases and recognized that the phase of the moon under which each of us is born is an accurate indicator of both our core personality type and our life purpose. He came to this conclusion through a close study of the interaction between the solar outer consciousness and the instinctual nature of lunar awareness.[1]

It follows naturally that when the moon transits its natal position, we become highly sensitized to our purpose and are probably more focused on this than at any other time in the month. It is also a time when specific challenges associated with our particular phase of the creative cycle will arise, giving us the opportunity to overcome them in the most propitious way possible.

In the 1950s, Czech psychiatrist Dr. Eugene Jonas also discovered that it was not uncommon for a woman to be most fertile during the lunation phase of her natal chart, often causing her to experience spontaneous ovulation. This accounts for why some women become pregnant at times outside their menstrual midpoint, an important factor to be taken into account both in terms of infertility and the efficacy of the natural methods of contraception.[2]

WHAT PHASE WAS I BORN IN?

To calculate your own lunation phase, identify the sign and degree of your sun and moon within your natal chart. As an example, let's use the following settings: the sun at 6 degrees Virgo and the moon at 4 degrees Cancer.

Starting at the sun, walk around the astrology chart counter-clockwise, counting the number of degrees separating the sun from the moon, remembering that each sign consists of 30 degrees. In this case, the degree difference is 298.

As summarized in chapter 2, there are eight different phases of the

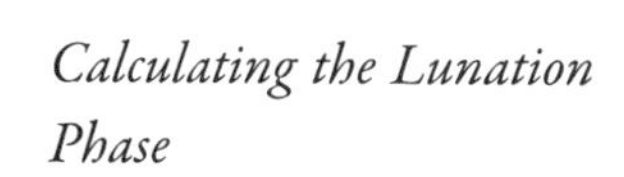
Calculating the Lunation Phase

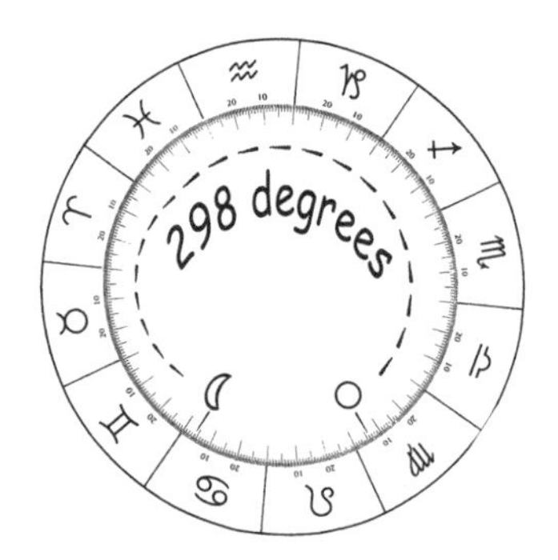

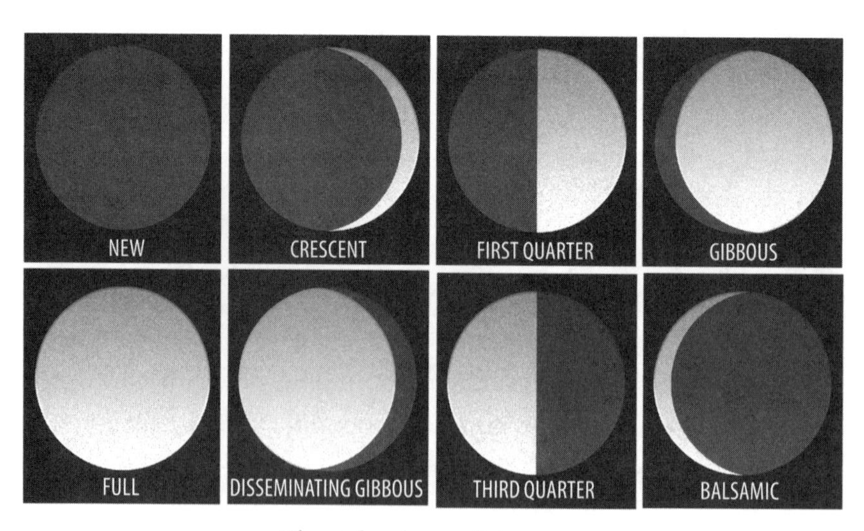

The eight phases of the moon

moon. By reading about your lunar phase, you will gain insight into your strengths and challenges as well as your all-important life purpose.

NEW MOON PHASE

Time: up to 3½ days after the new moon

Degree: 0–45 degrees ahead of the sun

Life Purpose: to germinate and emerge

Insight has been received and the seed of an idea has been planted but it is still germinating in the dark.

This phase occurs as the beginning of a whole new cycle of creativity. Those born into this lunation phase often have little learned awareness from the past, causing them to work best through spontaneity and instinct. The forces that drive them often arise from within and hence they may appear irrational to others, although their innocence and childlike enthusiasm is charming and irresistible.

Often without conscious awareness of their purpose and direction but trusting their instincts, they can feel disconnected from the earth. They are helped by becoming more present and aware of their bodies and by making a mark upon the world without seeking the approval of others to give them security.

CRESCENT MOON PHASE

Time: 3½–7 days after the new moon

Degree: 45–90 degrees ahead of sun

Life Purpose: to move and focus

The seedling now forces itself toward the light, challenged by an opposing force of gravity to stay within the comfort of the darkness.

Individuals born within this phase will find that as they strive to express their uniqueness, there emerge limiting beliefs and fears from past lives and/or from childhood, creating tensions that may cause them to fall back into the relative comfort of failure and inertia. In addition, they will be more susceptible to the emotional manipulation of others due to their poorly developed self-awareness. Therefore, a thickening of the stem is required to overcome ancestral and personal fears and insecurities.

With determined focus, however, there is excitement at the prospect of newness, which should not be lost when setbacks occur; these should be seen less as signs to withdraw and more as tools to strengthen the will. Through time, self-reliance develops as these individuals' identity starts to take root.

FIRST QUARTER PHASE

Time: 7–10½ days after the new moon

Degree: 90–135 degrees ahead of sun

Life Purpose: to build and decide

Now there is a strong determination to define goals and seek self-individuation.

With the moon square to the sun, the tension increases, actively compelling individuals born within this lunation phase to choose to follow their own soul's path and to build structures that nurture and protect their growth. Many challenges will be met as the old patterns are presented—and these are to be cleared once and for all. To do this, however, they must make choices and decisions from a clear and objective standpoint: from the stance of the hero rather than from the place of the victim.

Commonly, individuals born in this phase find that spiritual growth occurs through drama, conflict, and stress. They may even attract other peoples' problems into their lives or find themselves working in crisis situations. They are passionate and easily fired up by a challenge but need to learn how to manage stress skillfully through making decisions wisely and managing constructively rather than contributing to the chaos.

As they learn to manage their own energies and needs, a healthy realization of their dreams and ideas continues.

GIBBOUS MOON PHASE

Time: 10½–14 days after the new moon

Degree: 135–180 degrees ahead of sun

Life Purpose: to improve and perfect

As the moon moves farther from the sun, self-actualization increases. There is an impetus to assess and refine the situation to produce the perfect bloom.

Those born within this lunation phase are gifted with the potential for great success but may be handicapped by self-doubt and self-criticism. Such individuals often feel they are "never good enough" and have difficulty breaking free of the rigid views they have of themselves and others.

Always attempting to "get it right," they find themselves overanalyzing situations rather than using that same keen insight to access, refine, and move ahead. Old patterns of resistance surface in order to be acknowledged and transformed; flexibility and objectivity are the keys to success. At this point, the light is much brighter, which may be both a blessing and a fright, for there are now few places to hide and more pressure for them to fulfill their destiny.

FULL MOON PHASE

Time: 15–18½ days after the new moon

Degree: 180–225 degrees ahead of sun

Life purpose: conscious integration and success

Now the sun and moon oppose each other; the sun represents the outer expressed consciousness and the moon the instinctual nature.

Those individuals born during this phase soon learn that both poles of existence are important—the soul and the personality (the outer and inner worlds)—and that success is achieved when both are honored and respected. Without conscious acceptance of both, conflict arises, preventing either aspect from reaching fulfillment. Together, they can become highly successful; separate they will remain shadows.

For these individuals, this same dynamic plays out in relationships: Outwardly, they seek the perfect match or soul mate who will make life worth living. If such a mate is not found, they will direct their attention to a guru, teacher, or group in order to become complete. Eventually, it is hoped, they will see the light and realize that the sacred marriage they seek is not to be found in another but instead involves the synthesis of their own inner polarities, especially those between thought and action, spirit and matter.

As they consciously accept and integrate their polarities, their intention becomes embodied in reality and their life becomes full.

DISSEMINATING GIBBOUS MOON PHASE

Time: 3½–7 days after the full moon

Degree: 225–270 degrees ahead of sun

Life Purpose: to distribute and convey

After standing in its own light and knowing what it is like to be fully alive, the moon returns to the sun.

For those born within this phase, this is a time not only to enjoy the fruits of their endeavors, but also to offer others the insights they have gained. Thus, this is a life of teaching and sharing the wisdom gleaned from experience while always remembering that the greatest teacher is embodied expression. It is essential that those with this placement walk their talk. If they do not, the message is empty and the words have no meaning, leading to the lack of an audience and discontentment on the part of the messenger. At times like this, there is a need to regroup and reconnect to the inner teacher and ask: Do I believe in my own message? Another problem that can emerge is fanaticism: The mission to convey information becomes personal and objectivity is lost.

Ultimately, these individuals will find the contentment they seek through their willingness to engage, without judgment, the ideas and thoughts of others on collective social issues.

LAST OR THIRD QUARTER MOON PHASE

Time: 7–10½ days after the full moon

Degree: 270–315 degrees ahead of sun

Life Purpose: to revise and reevaluate

As the moon heads back toward the sun rapidly, the light of externalization starts to fade and the process turns inward.

Those individuals born within this lunation phase will find themselves constantly examining their beliefs in light of their experiences and asking: What is true? This mental scrutiny often sparks a crisis in consciousness, which can lead to several soul-searching events during their life. Their inner urges will force them to disengage from belief systems that are redundant, despite the coercion of others who have become dependent on the old, familiar ways.

During this fermentation process, these individuals can be thrown into inner turmoil as the old dies and there is no obvious successor; they are unable to share with others for fear of inciting disappointment and judgment. They wear masks, often unaware that the profound changes occurring in their own thought processes will affect not only their lives but the lives of generations to come, stretching far beyond their immediate family.

To survive this metamorphosis, it is vital not to throw the baby out with the bath water. Instead, they must value what has been learned and refined through lifetimes and remember that this will very soon become compost for the future seeds of inspiration.

BALSAMIC MOON PHASE

Time: 10½ days after full moon until the new moon

Degree: 315–360 degrees ahead of sun

Life Purpose: to distill and transform

The last light of the inverted crescent moon eventually fades and darkness falls as the developed seed containing the distilled essence of experience falls from the parent plant and lies dormant in the earth, awaiting germination.

Those born within this dark lunation phase will find their life embroiled with mystery, death, healing powers, and transformation. It is here that the moon goddess rejoins her sun god to give birth to their destiny. For these individuals, the world straddles the old and the new. Much effort is spent completing unresolved issues, which often leads to intense, short-lived relationships. From an early point on, they feel like outsiders to their family and peers, unable to connect to the ways of the past and unaware of any who speak their language as pioneers of new ideals.

Their inner sense of purpose causes them to focus all their attention on their prophetic visions so that their life will not be wasted; they often seek out a student, their son and heir, to carry forward their ideas even at their own personal sacrifice.

This is one of the most complex lunation phases and completes the creative cycle until once more the new seed is ready for germination and the new moon starts to form.

THE NODES OF THE MOON

Next we will discuss the moon's nodes as they relate to us as individuals. The South Node is like an old friend whose familiarity we seek, especially at times of stress. Within it, we find attitudinal patterns that are instinctual but not necessarily healthy, commonly hampering soul growth and keeping us in a rut. Enter the North Node, which reveals the direction we need to take in order to expand consciousness and experience fulfillment. It demands that we evolve beyond behaviors that can be addictive because of the comfort they offer.

It is not to be assumed that the south node is negative but rather that its gifts are well worn. As I say: "Just because you do something well, doesn't mean you should keep doing it." The gifts of the north node are different. They are so new that we may see them merely as a dream, not immediately recognizing that they belong to us. To fully embody the qualities of north node demands courage and the realization that the risk is worth taking.

Most astrologers see the south node as an indicator of past lives, commonly, the last life. It is interesting to enter a meditative state and allow oneself to imagine the focus of one's last life. This insight offers a good indication of the beliefs and challenges of the south node and what needs to be left behind in this life. To enhance our awareness, in one's early life the soul will often draw to it situations and experiences that reinforce the south node messages, so that we can use these influences to strengthen our resolve to move forward. Of course, it is possible to fall asleep to the urges from the north node and settle into a life fully directed by one's south node.

Yet, the pressure from these two nodes to move us forward never slackens. The south node acts as the bow and the north as the arrow, clearly directed at what we can become. They are aided by the Virgin and Crone, the former acting as an intuitive guide urging us forward, the latter destroying old beliefs that would trap us in karmic patterns.

Every 18.6 years, (a full Saros cycle), the nodes return to their original placement in our natal chart, increasing the north node's pressure to honor its gifts. This certainly happened in my life. When I was eighteen, my father died, forcing me to listen to my Capricorn north node and become self-responsible and disciplined and release my attachment to a very comfortable childhood. At thirty-eight, I was once again thrust out on my own after the breakdown of my first marriage while, at the same time, my first book was published and I started to carve a new path for my professional life.

The following information will give some insights into your own karma. Remember, the placement of your south node indicates your past karma and what you are leaving behind. The north node represents your soul's purpose in this lifetime and beyond.[3]

ARIES NORTH NODE, LIBRA SOUTH NODE

Develop: independence, self-awareness, trusting your own impulses, courage, and the willingness to take risks

Leave behind: selflessness; being too nice; seeking the opinions of others; obsession with fairness; codependency; and the need to find an ideal, committed partnership

TAURUS NORTH NODE, SCORPIO SOUTH NODE

Develop: loyalty, boundaries, patience, kindness, the ability to value your own gifts and talents, forgiveness, and an enjoyment of the senses

Leave behind: attraction to crisis situations, impatience, other people's business, intensity, judgmental nature, overreacting, and brooding

GEMINI NORTH NODE, SAGITTARIUS SOUTH NODE

Develop: curiosity, tact, logical thinking, asking what others think rather than assuming to know, seeing both sides of a situation, detachment

Leave behind: the need to be right, impatience, thinking you know what people are thinking without listening, careless spontaneity, taking shortcuts, and restlessness

CANCER NORTH NODE, CAPRICORN SOUTH NODE

Develop: noticing and validating feelings, empathy, nurturing self, humility, and allowing others in

Leave behind: the need to be in control of everything, an excessive focus on goals, overexcessive responsibility for others, thinking things have to be difficult, and hiding feelings

LEO NORTH NODE, AQUARIUS SOUTH NODE

Develop: individuality, a willingness to take center stage, following the heart's desires, strengthened willpower, enjoying life, and allowing the inner child to play

Leave behind: detachment, aloofness, waiting for others to prompt actions, waiting for knowledge before moving, and the need always to be different or the rebel

VIRGO NORTH NODE, PISCES SOUTH NODE

Develop: participation, bringing order from chaos, the creation of routines, setting boundaries, self-contemplation, service to others, and the enjoyment of beauty

Leave behind: being a victim, confusion, avoidance of making plans, escapism—especially into addictions and obsessions, daydreaming, self-doubt, and self-delusion

LIBRA NORTH NODE, ARIES SOUTH NODE

Develop: cooperation, diplomacy, awareness of others' needs, win-win situations, sharing and selflessness

Leave behind: impulsiveness, self-assertion, poor judgment, self-centeredness, and outbursts of anger

SCORPIO NORTH NODE, TAURUS SOUTH NODE

Develop: self-discipline, an interest in change and transformation, elimination of useless possessions, deep relationships, deep emotions, and sensuality

Leave behind: Over attachment to comfort, possessiveness, concern with ownership, stubbornness, and an obsession with sensual pleasures

SAGITTARIUS NORTH NODE, GEMINI SOUTH NODE

Develop: a reliance on intuition, speaking from higher consciousness, direct communication without censorship, patience, and a trust of self

Leave behind: second guessing, indecisiveness, always wanting more information, saying what others want to hear, gossiping, and superficiality

CAPRICORN NORTH NODE, CANCER SOUTH NODE

Develop: self-discipline, self-care, self-respect, loyalty, the ability to stay goal-orientated, basing actions on reason rather than emotions, honoring successes, and self-responsibility

Leave behind: dependency, moodiness, insecurity leading to inaction, limiting self through fear, avoidance of personal risk, and control of others through emotional overplay

AQUARIUS NORTH NODE, LEO SOUTH NODE

Develop: objectivity, desire for friendship, making decisions for the group, willingness to be unconventional, and group participation

Leave behind: an insistence on getting your own way, willfulness, attachment to the need for approval, a need to be the center of attention, and melodramatic tendencies

PISCES NORTH NODE, VIRGO SOUTH NODE

Develop: a nonjudgmental approach, compassion, a greater focus on the spiritual path, connection to higher states of consciousness, and expanded creativity

Leave behind: anxiety reactions, over analysis, a need for details and cleanliness, perfectionism, inflexibility, and faultfinding

CONCLUSION

THE HEART OF THE GREAT MOTHER

I hope you have enjoyed my teachings in this book as much as I have benefited from receiving them and sharing them with you. Here, I share one meditation I experienced that aptly sums up my feelings for and about the Great Mother.

The perspiration ran down my back in tiny rivers as I sat in the darkness, entranced by the rhythmic tones of the drums. I was in a sweat lodge, or *inipi,* following one of the most sacred ceremonies of the Native American tradition. A low dome-shaped structure built from natural materials, the lodge is seen to represent a womb offering new beginnings and insights to those who enter with reverence and humility.

Inside the inipi it was dark, apart from the glow emanating from the rocks, or grandfathers, that had been placed with great care and attention in the central pit. These spirits of the stone kingdom had been selected specifically by the medicine man who led our ceremony; he respected their wisdom and ability to inspire our prayers and meditations. As water was poured onto them, the temperature in the lodge increased. I was instantly enveloped by the breath of these great beings and urged to focus my awareness more deeply inward.

Now was the time to speak my prayers aloud, allowing the smoke and steam to carry them beyond the inipi to Wakan Takan, the Great Spirit. As I came to the end of my appreciation for those who had lived upon the earth and those who had passed into the world of spirit, I found myself sending loving thoughts to the star people. This surprised me because, although I respected my extraterrestrial connections, they did not usually feature in my prayers.

Such associations extended back into my early childhood, when a tunnel of light would lightly touch the top of my head as I lay in my cot at night. I felt no fear; rather, I sensed the opening of a portal to a world I knew beyond this earth plane. With joy, I would find myself entering the tunnel and merging immediately with a field of energy—a kind of journey that felt like coming home. When I was old enough to tell my mother about these nightly visits, she said wisely: "I don't know where you are going, but if you feel joy, follow your heart and you will find your path." It was many years later that a medium helped me to understand these energetic travels: "You do know you come from the stars?" At that moment, for the first time in my life I felt I'd been seen, with my heart knowing what my head had struggled to understand.

Since then, my connection to the star people had remained constant, with no need for greater explanation. Yet here I was sending prayers out into the galaxy, and I was amazed to hear the medicine man saying soon after: "There is a very old spirit from the stars who wants to join this ceremony. I will welcome him with song—and then hold onto your hats, for we are going on a celestial journey."

With that, more heated stones were added to the pit, the door was closed, and the songs began. Immediately, I found myself moving rapidly skyward accompanied by two dolphins, who I have always considered interdimensional travelers. As we left the earth's atmosphere, I looked back and was surprised to see a vibrant mass of brilliant blue energy emanating from the planet's surface. This was unlike the pictures taken by astronauts; I knew I was viewing the energetic aura of Gaia.

Our journey ended in front of a long table with nine chairs. Eight were positioned on either side of the table, facing each other, and one was one placed at the head of the table. It was clear that I was to sit in this seat, and as I did, from beneath my feet to above my head, I was immediately flooded with all the energy from the other eight places until my body was electric. At that moment I knew I had entered and embodied the ninth dimension associated with the Galactic Center and the Great Mother. My heart was resonating with her heart around which billions of stars and planets orbited in perfect harmony.

Through my heart, I heard a voice speak to me:

> I am the heart of the Great Mother, the perpetual fire of the galaxy. I am toroidal in shape, and it is through me that the consciousness of every star, planet, and life-form achieves transformation. It is through me that the essential energy of spirit passes into matter and matter transforms eventually back into my ocean of unlimited possibilities. Each star you see equates to the heart of an individual physically setting the pulse or rhythm for every cell of the body to follow. If the star is the heart, then the planets represent the organs, and the life forms living upon the planets are the cells.
>
> Inherent in all human beings is a signature frequency specific to a particular star system often carried within their skin color, language, songs, physical shape, or unique connection to the land. In the same way, the nature kingdoms resonate with the stars through their shapes, colors, and, in the case of birds, their songs. Unfortunately, over time, humanity has disconnected not only from the natural world but also from the built-in signal calling upon them to remember their galactic origins. Yet on a clear, starlit night, it is not uncommon to find yourself looking out at the stars and wondering why certain constellations seem to attract your attention.
>
> The great shift in consciousness that is occurring on planet Earth at this time is not exclusive to the human race. Such transformation affects us all, including me, the Great Mother. It requires us to recall and integrate all parts of our creative self and build the light body that will transport us to the next level. My heart is entraining the hearts of the stars, while the stars are calling on their children, humanity, to remember. But I offer a note of clarification: It is not enough to

> believe or even know you come from the stars, for it is the creative gems of your life experiences that will feed and nurture the consciousness of your celestial Mother. It is essential that all individuals connect first with their body, then their beliefs, and finally, their soul, clearing away the superfluous matter until there is only pure light.

Suddenly, my life's work made sense. As a teacher of mind-body medicine, I had encouraged participants and clients to listen to the inherent wisdom of the body, recognizing that the heart of each cell is programmed to resonate with the heart of the body. I employed body dialogue techniques to remind individuals that their body loves them, often to a far greater degree than they love themselves.

Then, I had added another dimension: assisting individuals to call into their heart all those parts of the self that had become separated through shame, fear, or limiting beliefs. Such psychological complexes, often leading autonomous lives while directing our actions, could be thousands of years old and yet now were being called home by the love of the Galactic Mother. All this teaching was wrapped in the knowledge that there was a small voice within everybody waiting to be heard—the intuition that acts as navigator of the soul.

I knew my time with the Great Mother was coming to a close and asked one final question: "What words of advice do you offer as we approach 2012 and beyond?"

Through my heart, I heard the answer:

> Still your mind and center in your heart. Now, through waves of love, connect to the hearts of your cells, organs, and chakras, remembering that in the heart there is only now and the promise of unlimited potential. Finally, extend your energy toward the stars through the embodiment of love for yourself, others, and the source of all existence, for essentially, there is only love, and love is enough.

And with that, I found myself journeying swiftly once again toward the earth until I felt the cool ground under my body and I was wrapped in the darkness of the inipi. The songs were ending, the ceremony was almost complete, and yet the next stage of my journey had just begun. As we stepped out into the cool night air, we looked up to see the Milky Way snaking across the sky above. My heart reached out to all children of the stars with the hope that each would hear the call of the Great Mother and feel the embrace of her unbounded love.

NOTES

INTRODUCTION. TEACHINGS BENEATH THE STARS

1. John Major Jenkins, *Maya Cosmogenesis 2012* (Rochester, Vt.: Bear & Co., 1998), 113–14.
2. Ibid., 10–11.
3. Ibid., 31.
4. Carlos Barrios, *Kam Wuj, El Libro del Destino* [The Book of Destiny] (Buenos Aires: South American Editorial, 2000).
5. Ibid.
6. William Henry, *Oracle of the Illuminati* (Kempton, Ill.: Adventures Unlimited Press, 2005), 188–90.
7. Carlos Barrios, *Kam Wuj, El Libro del Destino* [The Book of Destiny] (Buenos Aires: South American Editorial, 2000).
8. Ibid.
9. John Major Jenkins, *Maya Cosmogenesis 2012* (Rochester, Vt.: Bear & Co., 1998), 31.
10. Ibid., 52.
11. Ibid., 25.
12. Michio Kaku, *Hyperspace* (New York: Anchor Books, 1994), 16.
13. Mark Comings, *The Essential Nature of Space, Time and Light Consciousness,* presented at the International Institute of Integral Human Sciences conference, Montreal, 2005.
14. Dennis William Hauk, *The Emerald Tablet* (New York: Penguin Books, 1999), 45.
15. Tom Kenyon and Judi Sion, *The Magdalen Manuscript* (Orcas, Wash.: ORB Communications, 2002), 115–31.

16. Mark and Elizabeth Clare Prophet, *St. Germain on Alchemy* (Corwin Springs, Mont.: Summit University Press, 1993), 12.

CHAPTER 1. THE CREATION MYTH

1. Dennis Tedlock, *The Popul Vuh: The Definitive Edition of the Mayan Book of the Dawn of Life and the Glories of the Gods and Kings* (New York: Simon and Schuster, 1996).
2. James Robinson, ed., *The Nag Hammadi Library in English* (San Francisco: HarperCollins, 1990), 131, The Gospel of Thomas, verse 39.
3. Wikipedia, *The Puranas,* www.wikipedia.com.
4. Genesis 2:17.
5. Anne Llewellyn Barstow, *Witchcraze: A New History of the European Witch Hunts* (San Francisco: Pandora, 1994), 23.
6. Kathy Jones, *The Ancient British Goddess* (Glastonbury, UK: Ariadne Publications, 2001), 125.
7. Gerald Murphy, *The Iroquois Constitution,* The Cleveland Free-Net—aa300, distributed by the Cybercasting Services Division of the National Public Telecomputing Network (NPTN).

CHAPTER 2. RHYTHMS OF THE MOON

1. John Major Jenkins, *Maya Cosmogenesis 2012* (Rochester, Vt.: Bear & Co., 1998), 205.
2. Zecharia Sitchin, *The 12th Planet* (Rochester, Vt.: Bear & Co., 1991).
3. Barbara Marciniak, *Bringers of the Dawn* (Rochester, Vt.: Bear & Co., 1992), 17.
4. Laurence Gardner, *Genesis and the Grail Kings* (Gloucester, Mass.: Fair Wind Press, 2002), 246–50.
5. Ibid., 181.
6. Exodus 16:31.
7. Rick Strassman, *DMT: The Spirit Molecule* (Rochester, Vt.: Park Street Press, 2000), 56.
8. Lennart Möller, *The Exodus Case: New Discoveries Confirm the Historical Exodus* (Copenhagen: Scandinavia Publishing House, 2000).
9. Osman Ahmed, *Moses and Akhenaten: The Secret History of Egypt at the Time of the Exodus* (Rochester, Vt.: Bear & Co., 2002).

10. Dennis William Hauk, *The Emerald Tablet* (New York: Penguin Books, 1999), 25.

CHAPTER 3. THE HERO'S JOURNEY

1. Joseph Campbell, *The Hero with a Thousand Faces* (New York: Bollingen Foundation, 1949).

CHAPTER 4. THE CELESTIAL DESIGN

1. Alice Bailey, *Esoteric Astrology* (London: Lucis Press Ltd., 1951), 144.

CHAPTER 6. FROM BOY-CHILD TO KING

1. Alice Bailey, *Esoteric Healing* (London: Lucis Press Ltd., 1953), 183–87.
2. Christopher Knight and Robert Lomas, *The Hiram Key* (London: Element Books, 1997), 102.
3. Alice Bailey, *Esoteric Healing* (London: Lucis Press Ltd., 1953), 183–87.
4. Christopher Knight and Robert Lomas, *The Hiram Key* (London: Element Books, 1997), 105.
5. Ibid., 25
6. Ibid., 303
7. Diane Wolkstein and Samuel Noah Kramer, *Inanna, Queen of Heaven and Earth* (New York: Harper, 1983), 4–9.
8. C. G. Jung, *Psychology and Alchemy* (New York: Bollingen Foundation, 1968), 228.

CHAPTER 7. AND THEN ONE DAY . . .

1. Diane Wolkstein and Samuel Noah Kramer, *Inanna, Queen of Heaven and Earth* (New York: Harper, 1983), 52–57.
2. Eve Ensler, *The Vagina Monologues* (New York: Random House, 1998).

CHAPTER 8. THE LOVE THAT KNOWS NO END

1. Charlene Spretnak, *Lost Goddesses of Early Greece: A Collection of Pre-Hellenic Myths* (Boston: Beacon Press, 1981).

CHAPTER 9. THE DESCENT

1. Diane Wolkstein and Samuel Noah Kramer, *Inanna, Queen of Heaven and Earth* (New York: Harper, 1983), 52–60.
2. Laurence Gardner, *Genesis and the Grail Kings* (Gloucester, Mass.: Fair Wind Press, 2002), 160.
3. Kathy Jones, *The Ancient British Goddess* (Glastonbury, UK: Ariadne Publications, 2001), 46–47.
4. Dennis William Hauk, *The Emerald Tablet* (New York: Penguin Books, 1999), 231.
5. Tom Kenyon and Judi Sion, *The Magdalen Manuscript* (Orcas, Wash.: ORB Communications, 2002), 20.
6. Hans Jenny, *Cymatics* (Newmarket, N.H.: Macromedia, 2001).
7. Masaru Emoto, *The Message from Water* (Leido, The Netherlands: Hado Publishing, 1999).
8. Diane Wolkstein and Samuel Noah Kramer, *Inanna, Queen of Heaven and Earth* (New York: Harper, 1983), 62–67.
9. Thich Nhat Hanh, *Call Me by My True Names* (Berkeley: Parallel Press, 1999), 72.

CHAPTER 10. THE TRUTH SHALL SET THEM FREE

1. Hazrat Inayat Khan, *Teachings of Hazrat Inayat Khan; Purpose of life.* www.wahiduddin.net, vol. 1, chapter 8.
2. Joseph Chilton Pearce, *The Biology of Transcendence* (Rochester, Vt: Park Street Press, 2002), 57.
3. James Twyman, *Emissary of Love* (Findhorn, UK: Findhorn Press, 2002), 48.

CHAPTER 11. THE EMERALD TABLET

1. Mark and Elizabeth Clare Prophet, *St. Germain on Alchemy* (Corwin Springs, Mont.: Summit University Press, 1993), 6.
2. Dennis William Hauk, *The Emerald Tablet* (New York: Penguin Books, 1999), 32.
3. M. Doreal, *Hermes Trismegiste* (Paris: Messrs Firmin Didot 1858).

4. M. Doreal, *The Emerald Tablets of Thoth, the Atlantean* (Nashville: Source Books and Sacred Spaces, 1996).
5. Dennis William Hauk, *The Emerald Tablet* (New York: Penguin Books, 1999), 22.
6. Ibid., 25–30.
7. Ibid., 3.
8. Ibid., 45.
9. Ibid., 165.

CHAPTER 12. THE LUNATION PHASES AND THE NODES OF THE MOON

1. Dane Rudhyar, *The Lunation Cycle* (Santa Fe: Aurora Press, 1978).
2. Demetra George, *Finding Our Way through the Dark* (San Diego: ACS Publications, 1994), 15.
3. Jan Spiller, *Astrology for the Soul* (New York: Bantam Books, 1997).

BIBLIOGRAPHY

Ahmed, Osman. *Moses and Akhenaten: The Secret History of Egypt at the Time of the Exodus.* Rochester, Vt.: Bear & Co., 2002.

Bailey, Alice. *Esoteric Astrology.* London: Lucis Press Ltd., 1951.

———. *Esoteric Healing.* London: Lucis Press Ltd., 1953.

Barrios, Carlos. *Kam Wuj, El Libro del Destino* (The Book of Destiny). Buenos Aires: South American Editorial, 2000.

Campbell, Joseph. *The Hero with a Thousand Faces.* New York: Bollingen Foundation, 1949.

Doreal, M. *Hermes Trismegiste.* Paris: Messrs Firmin Didot, 1858.

———. *The Emerald Tablets of Thoth, the Atlantean.* Nashville, Tenn.: Source Books and Sacred Spaces, 1996.

Emoto, Masaru. *The Message from Water.* Tokyo: Hado Publishing, 1999.

Ensler, Eve. *The Vagina Monologues.* New York: Random House, 1998.

Gardner, Laurence. *Genesis and the Grail Kings.* Gloucester, Mass.: Fair Wind Press, 2002.

George, Demetra. *Finding Our Way through the Dark.* San Diego: ACS Publications, 1994.

Hauk, Dennis William. *The Emerald Tablet.* New York: Penguin Books, 1999.

Henry, William. *Oracle of the Illuminati.* Kempton, Ill.: Adventures Unlimited Press, 2005.

Jenkins, John Major. *Maya Cosmogenesis 2012.* Rochester, Vt: Bear & Co., 1998.

Jenny, Hans. *Cymatics.* Newmarket, N.H.: Macromedia, 2001.

Jones, Kathy. *The Ancient British Goddess.* Glastonbury, UK: Ariadne Publications, 2001.

Jung, C. G. *Psychology and Alchemy*. New York: Bollingen Foundation, 1968.

Kaku, Michio. *Hyperspace*. New York: Anchor Books, 1994.

Kelley, David H. "Mesoamerican Astronomy and the Maya Calendar Correlation Problem," *Memorias del Segundo Coloquio Internacional de Mayistas* 1 (1989).

Kenyon, Tom, and Judi Sion. *The Magdalen Manuscript*. Orcas, Wash.: ORB Communications, 2002.

Khan, Hazrat Inayat. *Teachings of Hazrat Inayat Khan; Purpose of life*. www.wahiduddin.net, vol. 1, chapter 8.

Kinstler, Clysta. *The Moon Under Her Feet*. San Francisco: HarperCollins, 1991.

Knight, Christopher, and Robert Lomas. *The Hiram Key*. London: Element Books, 1997.

Marciniak, Barbara. *Bringers of the Dawn*. Rochester, Vt: Bear & Co., 1992.

Möller, Lennart. *The Exodus Case: New Discoveries Confirm the Historical Exodus*. Copenhagen: Scandinavia Publishing House, 2000.

Nhat Hanh, Thich. *Call Me by My True Names*. Berkeley: Parallel Press, 1999.

Pearce, Joseph Chilton. *The Biology of Transcendence*. Rochester, Vt: Park Street Press, 2002.

Prophet, Mark, and Elizabeth Clare Prophet. *St. Germain on Alchemy*. Corwin Springs, Mont.: Summit University Press, 1993.

Robinson, James, ed. *The Nag Hammadi Library in English*. San Francisco: HarperCollins, 1990.

Rudhyar, Dane. *The Lunation Cycle*. Santa Fe: Aurora Press, 1978.

Sitchin, Zecharia. *The 12th Planet*. Rochester, Vt: Bear & Co., 1991.

Spiller, Jan. *Astrology for the Soul*. New York: Bantam Books, 1997.

Spretnak, Charlene. *Lost Goddesses of Early Greece: A Collection of Pre-Hellenic Myths*. Boston: Beacon Press, 1981.

Strassman, Rick. *DMT: The Spirit Molecule*. Rochester, Vt.: Park Street Press, 2000.

Tedlock, Dennis. *The Popul Vuh: The Definitive Edition of the Mayan Book of the Dawn of Life and the Glories of the Gods and Kings*. New York: Simon and Schuster, 1996.

Twyman, James. *Emissary of Love*. Findhorn, UK: Findhorn Press, 2002.

Walker, Barbara. *The Woman's Encyclopedia of Myths and Secrets*. San Francisco: Harper, 1983.

Wolkstein, Diane, and Samuel Noah Kramer. *Inanna, Queen of Heaven and Earth*. New York: Harper, 1983.

INDEX

2012, 3, 51, 218

age of
  Aries, 81
  Aquarius, 17, 79–80, 82, 83, 94
  Cancer, 82
  Gemini, 82
  Leo, 82
  Pisces, 17, 79–81
  Taurus, 81
ajna. *See* third eye chakra
akashic record. *See* blueprint
Akhenaten, 66–67, 191–92
alchemist, 9, 11, 13, 23, 32, 48–49, 66–67, 91, 98, 111, 188–89, 195, 199–201
alchemy, 13–14, 19, 53, 61, 65–67, 76, 86–87, 91, 98, 101, 103, 105, 110, 116, 119, 122, 136–37, 147–48, 152, 159, 174, 178, 187–90, 193–95, 202
amrita, 48–49, 52, 70, 73, 139, 166, 187. *See also* elixir of life
apple, 130, 165–67
Aquarius, 37, 76, 78–80, 82–83, 94, 186, 212, 214
archetype, 9, 16, 19–21, 23, 42–43, 54, 58, 60, 68, 72, 79–81, 84, 89, 94, 102, 108, 116, 137, 149, 153, 160, 168, 170–71, 173–74, 178, 184
Aries, 36, 38, 76, 78, 81, 96, 98, 119, 212–13
Artemis, 24, 40, 156
Asclepius, 105
astrological qualities
  cardinal, 76, 78–79, 98, 110, 137, 178
  fixed, 76, 78–79, 101, 116, 152, 187
  mutable, 76–77, 86, 103, 119, 174
astrology, 53, 76, 204. *See also specific astrological signs*
Athena. See Pallas Athena
Atlantis, 190
Aubrey holes, 55
Avebury, 90, 184

Ba, 13, 15, 49, 187, 201. *See also* higher self
Babylon, 61, 141, 163
Balinas, 192
base chakra, 96, 104, 134, 168, 177, 184
black bird, 159, 196–98
black hole, 3, 17, 64, 181, 185
Black Madonna, 135, 161. *See also* Mary Magdalene
Black Road, 2–3. *See also* Dark Rift, Great Cleft
blueprint, 13, 28, 32, 36, 75, 85, 87, 89, 92–93, 95, 98, 100, 116, 125, 141, 162, 169, 176, 182. *See also* higher self

Boaz, 104, 106, 108, 165, 167, 190, 202
boy-child. *See* puer
Brigid, 54, 89–92. *See also* St. Bride

caduceus, 105–6, 160
calcination, 98, 195–97
calendar, 3, 22, 51, 57, 90
Cancer, 36, 69, 76, 78, 82, 96, 110, 113, 204, 212–13
Capricorn, 36, 76, 78, 177–78, 211–13
cauldron, 14–15, 47, 53, 90, 142, 151, 155, 161, 163–64, 169, 199–200
celebrate, 14, 18, 41, 46, 70, 116–17, 122, 148–49
celestial soul. *See* Ba
Celtic, 54, 68, 90, 128–32, 163–64
Ceres. *See* Demeter
Cerridwen, 153, 163
chakra, 13, 38, 46, 48–49, 96, 104, 106, 119, 123, 134, 160, 162, 166, 168, 175–77, 184, 186, 193, 196–202, 218. *See also specific chakras*
Chang'e, 47
chaos, 3, 7, 21, 33–34, 69, 107, 122, 142, 152, 154, 206, 213
Christ consciousness, 83, 125–27
coagulation, 86, 148, 187, 201
cockerel, 199
conjunction, 110, 116, 196, 198
cow, 24–25, 67, 81
creator, 2, 15, 27, 64, 97, 115, 125, 127–28, 147
Crone, 14–16, 27, 29, 36, 48, 72, 87, 90, 96, 102, 110, 112, 119–20, 122–25, 128, 135, 138, 141, 151–53, 155–63, 165, 167, 169–72, 198–99, 202, 211
crop formation, 9–10, 58–59, 184
cross
  cardinal, 76, 78–79, 98, 110, 137, 178
  fixed, 76, 78–79, 101, 116, 152, 187
  mutable, 76–77, 86, 103, 119, 174
crow. *See* black bird
crown chakra, 49, 104, 106, 119, 123, 134, 176, 202
crystal children, 89, 183
Cunti, 129

Dark Goddess. *See* Crone
dark night of the soul, 122
Dark Rift, 2–3, 8, 151. *See also* Great Cleft; Black Road
Demeter, 23, 25–26, 138–40, 142–44, 149, 163
destroyer, 27, 72, 125, 128, 147, 153, 160
Diana, 24, 29
dissolution, 17–18, 36, 101, 182, 195, 197
distillation, 148, 174, 178, 186–87, 200–201
Djed, 13, 15, 49, 187
DMT, 65, 106, 154
DNA, 64, 184
dodecahedron, 5–6, 59
dolphin, 19, 88, 216
duality, 27, 33, 39, 45, 71, 73, 77, 86–87, 103–4, 110, 167, 183, 199

eagle, 112, 115, 157, 167
earthchild chakra, 168
Easter, 56
eclipse
  lunar, 51–53, 55
  solar, 51, 53
ego, 14, 18, 35–38, 73, 97, 99, 116, 135, 186, 194, 197, 200
Egypt, 13, 24–25, 46, 49, 58, 66–68, 81, 104–8, 145–47, 154, 158, 190–92
Eleusinian mysteries, 26, 139–40, 143
elixir of life, 21, 45, 47–48, 65–66, 112, 119, 161, 163, 165–66, 169, 186, 193, 201. *See also* amrita
Emerald Tablet, 67, 105, 188, 191–93
Enki, 118, 172
Ereshkigal, 118, 123–24, 150–51, 172, 177

eternal, 1, 3–4, 9, 11, 14, 19, 21, 23–24, 27, 39, 41, 45–48, 52, 54, 65, 72–74, 79, 84–85, 87–88, 90–91, 95, 104, 107–8, 125, 127, 143, 148, 151, 155–56, 160, 162–64, 167–68, 182–85, 187, 202. *See also* immortality
ether, 5–7, 17, 19, 30, 39, 48–49, 57, 59, 63, 89, 141, 183

Fibonacci's numbers, 37
fifth world, 3, 7, 17, 57, 84, 89, 183
Fisher King, 164
Freemason, 104, 108, 189

Galactic Center, 2–3, 9–11, 18–19, 50, 152, 181–82, 185, 217
Galahad, 164
Garden of Eden, 166–67
Gemini, 36, 76–77, 81–82, 96, 103–5, 212–13
Glastonbury, 29–31, 91, 134
god, 4, 19, 21–25, 33, 40, 45–46, 52, 57, 60–68, 70, 72, 74, 76, 81, 90, 98, 104–8, 118, 127, 130, 135, 138–41, 149, 154–56, 158–59, 162–63, 167–68, 172, 188, 191, 193, 210. *See also specific gods*
goddess, 4, 14–16, 19, 21, 24–29, 32–34, 36, 40, 45–48, 54, 63, 70, 72, 76, 81–82, 87, 90, 92–93, 107, 112, 119–20, 122, 124–26, 128–30, 132–35, 138–41, 144–45, 151–61, 163, 166, 169, 172, 178, 198–99, 210. *See also specific goddesses*
gold, 13, 18, 20, 25–26, 37, 48, 65, 67, 81, 90–91, 123, 138–39, 150, 169, 171, 185, 190, 197, 200
Golden Pill, 200
Great Art, 188, 199
Great Cleft, 2. *See also* Black Road; Dark Rift
Great Mother, 1–4, 6–12, 14–15, 17, 23, 25, 32, 35–37, 44, 47, 60, 63, 65, 75, 87, 96, 99, 100–101, 103–4, 110, 116, 119, 121, 127–28, 136, 138, 140, 146, 149, 151, 154, 161–62, 167–68, 172, 174, 177–78, 181–85, 187, 197, 215, 217–19. *See also* Mother
grid, 5–6, 58–60, 131, 185

Hades, 138, 140–41, 145. *See also* Pluto
Hathor, 24–25, 67, 81
heart, 6, 11, 12, 15, 18–19, 35, 39, 45, 59–60, 63, 75–76, 88, 91, 94–95, 99, 101, 103, 112, 119, 122, 124, 127, 142, 146, 153, 161, 164, 167, 169–71, 173, 176, 178, 180–86, 191, 195, 199–200, 212, 216–19
heart chakra, 177, 200–201
heart of Great Mother, 2–3, 12, 60, 65, 149, 151, 174, 181–82, 185, 187, 217
hearth, 54, 91, 178, 180
heaven, 5, 10, 16, 25, 36, 49–50, 62–65, 70, 74, 76–77, 81, 104, 118, 123, 127, 150, 156, 158, 183, 194, 200
Hecate, 29, 138, 141
hell, 57, 136, 155, 164
  Hel, goddess, 155
Hermes, 105, 140, 158, 189–92, 194, 202
hero, 14, 29, 36, 39, 41–42, 44, 68–70, 76, 103–4, 109–10, 114–15, 117, 131, 134, 137, 165, 186, 196–97, 206
higher self, 13, 75, 84, 86–87, 92, 121. *See also* Ba; blueprint
hologram, 6,11, 22–23, 59, 69, 85, 92, 171, 185
Holy Grail, 90, 164–65, 184
Horus, 25, 68, 146–48, 159

ida, 96, 104, 117, 183, 186

illusion, 2, 6, 16, 22–23, 72, 74, 113–14, 166, 196, 198–99
imagination, 23, 34, 85, 94, 189, 195, 197, 200. *See also* One Thing
immortality, 3–4, 9, 11, 13, 20–21, 26, 33, 38, 45–49, 52, 61, 70–73, 87, 108, 112, 130, 133, 139, 143, 148, 161, 166–67, 182, 187–88, 193. *See also* eternal
Inanna, 22, 109–10, 112–13, 118, 120, 122–24, 133, 141, 147, 150–51, 160, 166, 169, 172, 176–77
intention, 32–33, 59, 109–10, 123, 176, 196, 208
intuition, 7, 14, 28, 33, 91–93, 95, 103, 118, 156, 165, 172, 195, 198, 213, 218
Iroquois Constitution, 34. *See also* peace maker
Ishtar, 45, 141, 162
Isis, 24, 145–48, 153, 158–59, 161–62
Isle of Avalon, 30, 159

Jachin, 104, 106, 108, 165, 167, 190, 202
Jung, Karl, 93, 111, 170, 189

Ka, 12–15, 49, 57, 75, 106, 148, 155, 161, 168, 170, 187, 201. *See also* light body
Kali, 24, 29, 128, 133, 153, 158, 160, 162–63
Ketu, 52–53
Khat, 13
Kildare, 54–55, 90
king, 14, 16, 29, 36, 42–44, 46, 82, 96, 101–102, 108, 110, 114–17, 119–20, 122–23, 129–31, 135, 139, 147, 154, 156, 159, 167, 171, 178, 186, 190, 198
King Solomon, 107
Knights Templar, 108, 135
Kore, 138, 140–42, 144–45. *See also* Persephone
kundalini, 91, 133, 163

labyrinth, 29, 133–35
Lemuria, 92
Leo, 36, 38, 76, 78, 82, 96, 116, 119, 212, 214
Libra, 36, 76, 78, 137, 212–13
light body, 12, 13, 15–16, 33, 49, 148, 161, 168, 170, 201, 217
See also Ka
lightning rod, 9, 46, 59
Lilith, 48, 109, 112, 115, 176
lily, 48, 112
lion, 45, 82, 112, 116–17, 153–54, 167
lotus, 48, 112, 155
lover, 12, 29, 36, 43, 124–25, 137, 142, 164, 169, 186, 199
lunar, 40, 44, 47, 51–53, 55–57, 61, 65–67, 120, 133, 148, 203, 205
lunar standstills, 55–56, 133

Ma'at, 106–8, 189, 191, 202
magician, 4, 9, 12, 15, 37, 43, 96, 104, 117, 119, 160, 164–65, 169, 175, 186–87
Mah. *See* Mut
manna, 64–65, 67
Mary, 24, 29, 88–89, 135
Mary Magdalene, 135, 161
Maya, 2–3, 5, 8, 21–22, 51–52, 54
mentor, 102–3, 157
Milky Way, 2, 8, 20, 25, 151, 219
mirror, 14, 18, 21, 33, 36, 50, 91, 110, 113–14, 160, 172, 176, 197
moon, 3, 24, 39–67, 86–87, 90, 98, 101, 110, 116, 119, 133–34, 137–38, 144, 148, 152–53, 158, 175, 178, 187, 191, 193, 195–96, 202–14. *See also* lunar

moon phases
balsamic, 41, 43, 178, 187, 204, 209
crescent, 40–42, 46, 49, 57, 60–61, 101, 153,
191, 204–5, 209
disseminating gibbous, 41, 43, 119, 137, 204, 208
first quarter, 41–42, 103, 204, 206
full, 41, 43–44, 51–52, 55–56, 66, 116, 204, 207–9
gibbous, 41–42, 110, 204, 207
new, 3, 41–44, 51, 56, 86, 98, 204–7, 209–10
third quarter, 41, 43, 152, 175, 204, 209
moon tree, 54, 86
Moses, 25, 66–67, 81, 192
Mother, 23–27, 32–34, 36, 44, 46, 67, 71, 87, 96, 98–99, 101, 120, 124, 128–30, 132–33, 138, 140, 144–46, 149, 151, 155, 160–61, 163, 168, 170–71, 175, 178, 193, 195–96, 202, 216, 218
See also Great Mother
Mount Sinai, 66–67
Mut, 157–58

Nanna, 60–61, 67
Narcissus, 138, 142
nectar, 46, 48, 52, 70, 97
Nephilim, 63
Newgrange, 130–33
nodes of moon
north, 53, 210–14
south, 53, 210–14
See also Rahu and Ketu
nonlocal reality, 10–11, 61, 183
nurturer, 24, 27, 125

obelisk. *See* pillar
ocean of possibilities, 1, 4, 9, 11, 18, 23–24, 36, 50, 71, 84, 94–95, 98, 185, 197
One Mind, 2, 32–33, 35, 37, 71, 98, 191–93, 195–96, 200
One Thing, 23, 32–33, 35, 37, 71, 192–93, 195, 197, 200
Osiris, 46, 68, 145–48, 158
Ouroboros, 11–12, 38

Pallas Athena, 92–95, 114, 121
peace maker
Deganawidah, 34–35
peacock, 174, 199–200
Pele, 155, 157
pelican, 178, 201
perpetual fire, 54–55, 78, 91, 169, 177–80, 185, 200, 217
Persephone, 25, 138, 140–41, 143–44, 153
philosopher's stone, 67, 193, 201
phoenix, 202
pillar, 4, 104, 106–8, 111, 165, 167, 183, 190, 202. *See also* Jachin and Boaz
pineal gland, 48–49, 65, 106, 154–55, 160
pingala, 104, 119, 183, 186
Pisces, 36–37, 76, 78–81, 85–86, 94, 96, 104, 213–14
pituitary gland, 65, 160
Platonic Solids, 5–6
Pluto. *See* Hades
poles of existence, 6, 39, 45, 71, 110, 114, 125, 207
pomegranate, 140, 143–44
portal, 3, 8, 16, 22–23, 29, 58, 60, 79, 118–19, 124–25, 134–35, 159, 200, 216
precession of equinoxes, 79–80
puer, 36, 42, 90, 96, 98–99, 101, 103, 124, 131, 186
putrefaction and fermentation, 119, 137, 148, 152, 159, 161, 174, 199–200, 209

quantum plenum. *See* void

Ra, 25, 154
Rahu, 52–53, 71, 73
Rosslyn Chapel, 108

sacral chakra, 177
sacred geometry, 18, 38, 58, 120, 126–27, 181. *See also specific forms*
sacred marriage, 5, 13, 15, 39, 49, 111, 114, 162, 198, 201–2, 208
sacred sites, 18, 23, 31, 55–56, 58–60, 129, 131–32, 184. *See also specific sites*
sage, 36, 43, 102, 153, 169, 174–76, 186
Sagittarius, 36, 76–77, 174, 212–13
St. Bride, 29, 90
  mound of, 30–31
  church of, 91
St. Germain, 13, 188
Samudra Manthan, 25, 52, 70–71, 120. *See also* sea of milk
Saros Cycle, 54–56
Saturn, 100
Scorpio, 36, 38, 76, 151, 212–13
sea of milk, 23, 25, 70–71, 73. *See also* ocean of possibilities
Sehkmet, 82, 153
separation, 103, 196–97
serpent, 8, 11, 21, 26, 38, 52–53, 70–71, 91, 104–5, 109, 112, 114, 132–33, 162–64, 166–67, 186. *See also* Ouroboros
Set, 145–48
sexual fire, 13, 91, 112, 136, 162
Shakti, 27, 162
shaman, 8, 10–11, 16, 37, 123, 154, 159
shamrock, 129–30
Sheela-na-gigs, 127–29
Shiva, 70, 72, 158, 160
Snow White, 166
Sin, 45, 57, 61, 64, 66
Sitchin, Zecharia, 63
solar plexus, 177, 198
solstice
  summer, 56, 131, 133, 156
  winter, 56, 130–31, 166
Sophia, 162
spear of destiny, 164
Sphinx, 82–83, 154
Stonehenge, 55–56, 60, 90, 131
Sumer, 24, 60–61, 67, 109, 118, 141, 176
sushumna, 104, 119, 183, 186
swan, 105, 159–60, 199–200
sword, 115, 167, 196, 198
  of David, 165

Taurus, 36, 76, 78, 81, 96, 101, 212–13
third eye chakra, 105, 145, 160, 169, 176, 186, 201–2
thirty-six-year, 2–3
Thoth, 66, 105, 107, 146, 189–91
throat chakra, 177, 200
Tree of Life, 5, 108, 166–67, 184
Triple Goddess, 24, 27–29, 32, 36, 87, 119–20, 124–26, 129–30, 132, 138, 145, 151–52, 160, 163, 178
time-space, 10, 22
torus, 45, 180–85, 217
Tower of Babel, 61–62, 64
transformer, 2, 11, 47, 182
trinity, 27, 86–87, 105, 107–8, 129, 167, 202
turtle, 70, 72, 120
twenty-six thousand years, 2, 4, 15, 33, 73, 75, 79, 101, 187

underworld, 8, 26, 29, 31, 52, 57, 97, 102, 118–20, 131, 133, 135–36, 138, 140–42, 144–45, 147, 149–50, 155, 159–61, 165–66, 169, 172, 186, 190. *See also* Hell
unicorn, 45

unified field, 4–5, 39, 61, 73, 83, 86, 89, 127
unity, 6–8, 18, 45, 60, 81, 125, 127, 181, 183

Valkyries, 159
vesica piscis, 29, 35, 119–20, 125–27
Vesta, 178, 180
vestal virgins, 55, 178–80, 185
Virgin, 1, 27, 29, 36, 44, 54–55, 84, 86–90, 92, 95–99, 101–102, 110, 119, 121–22, 124, 128, 131, 135, 138, 156, 160, 169, 176, 178–80, 185, 202, 211. *See also specific Virgins*
Virgo, 36, 76–77, 119, 122, 204, 213–14
Vishnu, 52, 70–71, 161
void, 10, 17, 23, 120, 136
vulture, 16, 153, 157–58, 174
vulva. *See* yoni

wand, 4, 12, 15, 37, 96, 104, 117, 119, 160, 164, 186, 195
whale, 87–88
white hole, 3
wings, 19, 105, 112, 145, 157, 160, 169, 186, 198

Yahweh, 107
yin and yang, 86
yoni, 25, 29, 124–27, 128–30, 160

Zeus, 25, 93, 138, 140, 142, 145
ziggurat, 46, 61, 104. *See also* Tower of Babel
zodiac, 5, 36, 55, 76–78, 82, 148, 177. *See also* astrology

p. 183 Emissary of Love

BOOKS OF RELATED INTEREST

Maya Cosmogenesis 2012
The True Meaning of the Maya Calendar End-Date
by John Major Jenkins

Galactic Alignment
The Transformation of Consciousness According to Mayan, Egyptian, and Vedic Traditions
by John Major Jenkins

The Mayan Calendar and the Transformation of Consciousness
by Carl Johan Calleman, Ph.D.

The Mayan Code
Time Acceleration and Awakening the World Mind
by Barbara Hand Clow
Foreword by Carl Johan Calleman, Ph.D.

The Mystery of the Crystal Skulls
Unlocking the Secrets of the Past, Present, and Future
by Chris Morton and Ceri Louise Thomas

Return of the Children of Light
Incan and Mayan Prophecies for a New World
by Judith Bluestone Polich

The Mayan Factor
Path Beyond Technology
by José Argüelles

Time and the Technosphere
The Law of Time in Human Affairs
by José Argüelles

Inner Traditions • Bear & Company
P.O. Box 388
Rochester, VT 05767
1-800-246-8648
www.InnerTraditions.com

Or contact your local bookseller